Bacteriology: The Study of Bacteria

Bacteriology: The Study of Bacteria

Justin Clifford

SYRAWOOD
PUBLISHING HOUSE

New York

Published by Syrawood Publishing House,
750 Third Avenue, 9th Floor,
New York, NY 10017, USA
www.syrawoodpublishinghouse.com

Bacteriology: The Study of Bacteria
Justin Clifford

International Standard Book Number: 978-1-64740-093-4 (Hardback)

Cataloging-in-Publication Data

Bacteriology : the study of bacteria / Justin Clifford.
    p. cm.
Includes bibliographical references and index.
ISBN 978-1-64740-093-4
1. Bacteriology. 2. Bacteria. 3. Microbiology. I. Clifford, Justin.
QR74.8 .B33 2022
579.3--dc23

Table of Contents

Preface

This book aims to help a broader range of students by exploring a wide variety of significant topics related to this discipline. It will help students in achieving a higher level of understanding of the subject and excel in their respective fields. This book would not have been possible without the unwavered support of my senior professors who took out the time to provide me feedback and help me with the process. I would also like to thank my family for their patience and support.

Bacteria are prokaryotic microorganisms which are generally a few micrometers in length. The biological branch which deals with the studies related to the morphology, genetics, ecology and biochemistry of bacteria is referred to as bacteriology. It is a sub-field of microbiology that is particularly concerned with the identification, classification and characterization of bacterial species. Bacteriology is applied in the development of vaccines for various diseases such as diphtheria and tetanus. There are numerous areas where principles of bacteriology are applied and studied such as agriculture, marine biology and biotechnology. This book is compiled in such a manner, that it will provide in-depth knowledge about the theory and practice of bacteriology. The topics included herein on this topic are of utmost significance and bound to provide incredible insights to readers. This book aims to serve as a resource guide for students and experts alike and contribute to the growth of the discipline.

A brief overview of the book contents is provided below:

Chapter – Introduction

The branch of biology that focuses on the study of bacteria, as well as their ecology, morphology, genetics and biochemistry, is referred to as bacteriology. This is an introductory chapter which will introduce briefly all the aspects of bacteria and bacteriology such as the harmful effects of bacteria and their role in soil.

Chapter – Bacterial Taxonomy

The rank-based classification of bacteria is known as bacterial taxonomy. It also plays a vital role in the nomenclature of bacteria. The topics elaborated in this chapter will help in gaining a better perspective about bacterial taxonomy as well the different phyla of bacteria which are studied within this field.

Chapter – Gram Negative and Gram Positive Bacteria

Gram negative bacteria are the bacteria which do not retain the crystal violet stain which is used in the gram-staining method of bacterial differentiation. The bacteria that give a positive result in the Gram stain test are known as gram positive bacteria. The topics elaborated in this chapter will help in gaining a better perspective about the different types of gram negative and positive bacteria.

Chapter – Bacterial Biology

The shape, projections and cytoplasm of bacteria are studied within the domain of bacterial biology. It is also involved in the genetics of bacteria as well as their growth and multiplication. These diverse aspects of bacterial biology as well as the interactions of bacteria with other organisms have been thoroughly discussed in this chapter.

Chapter – Industrial Applications of Bacteria

Bacteria are used in various industries for diverse purposes such as food manufacturing production of probiotics, antibiotics, drugs, vaccines, insecticides, enzymes, fuels and solvents. The chapter closely examines these industrial applications of bacteria to provide a thorough understanding of the subject.

Justin Clifford

1

Introduction

The branch of biology that focuses on the study of bacteria, as well as their ecology, morphology, genetics and biochemistry, is referred to as bacteriology. This is an introductory chapter which will introduce briefly all the aspects of bacteria and bacteriology such as the harmful effects of bacteria and their role in soil.

BACTERIA

Bacteria (singular: bacterium) are a group of microscopic, single-celled prokaryotes—that is, organisms characterized by a lack of a nucleus or any other membrane-bound organelles.

Although among the most primitive organisms, bacteria reflect many universal features of life, including that they are composed of cells, transmit genetic information via DNA, and need energy from the environment to exist, grow, and reproduce; even sexual reproduction has been exhibited in some species of bacteria. Bacteria are often viewed negatively, given this group's connection to diseases. However, bacteria perform invaluable, beneficial functions in ecosystems, and also reflect harmony between living organisms in a number of ways. These include conversion of atmospheric nitrogen to forms that plants can use, exhibiting mutualism (a type of symbiosis in which both organisms in two interacting species receive benefit), and recycling nutrients through bacterial decomposition of dead plants and animals. Bacteria also provide an aid in digestion for many organisms, and are helpful in yogurt production, sewage treatment, and as sources of medicinal drugs.

Bacteria are the most abundant of all organisms. They are ubiquitous in both soil and water and as symbionts of other organisms. Many pathogens (disease-causing organisms) are bacteria. Most bacteria are minute, usually only 0.5-5.0 μm in their longest dimension, although giant bacteria like Thiomargarita namibiensis and Epulopiscium fishelsoni may grow past 0.5 mm in size. Bacteria generally have cell walls, like plant and fungal cells, but with a very different composition (peptidoglycans). Many move around using flagella, which are different in structure from the flagella of other groups.

Bacterial bodies may be spherical, rod-shaped, or spiral/curved shaped. Although unicellular, some bacteria form groupings of cells, such as clusters, filaments, or chains.

Bacteria: The Good, the Bad and the Ugly

The Good

The species of bacteria that colonize our respiratory and digestive systems help set up checks and balances in the immune system. White blood cells police the body, looking for infections, but they also limit the amount of bacteria that grow there. Likewise, bacteria keep white blood cells from using too much force. Bacteria also help out by doing things cells are ill-equipped to do. For instance, bacteria break down carbohydrates (sugars) and toxins, and they help us absorb the fatty acids which cells need to grow. Bacteria help protect the cells in your intestines from invading pathogens and also promote repair of damaged tissue. Most importantly, by having good bacteria in your body, bad bacteria don't get a chance to grow and cause disease.

The Bad

Some species of bacteria in your body can result in diseases, such as cancer, diabetes, cardiovascular disease, and obesity. Usually, these diseases happen only when the normal microbiome is disrupted, but that can occur even from antibiotics. Antibiotics kill bacteria, and some of those will be good bacteria that we need to protect our health. When that happens, the bad bacteria that normally are kept in check have room to grow, creating an environment ripe for disease.

Bad bacteria can exist at low levels in your body without causing harm or can grow too much and wreak havoc. Staphylococcus aureus can cause something as simple as a pimple or as serious as pneumonia or toxic shock syndrome. P. gingivalis can cause gum disease, and was recently linked to pancreatic cancer. Similarly, when not suppressed by good bacteria, Klebsiella pneumonia can cause colitis, and subsequently lead to colorectal cancer.

The Ugly

In addition to allowing disease-causing bacteria to flourish, the elimination of good bacteria throws the immune system out of whack. The result can be simple allergies or very debilitating autoimmune diseases. Without the right balance of bacteria, your body might suffer from constant inflammation.

Inflammation is the body's alarm system, which calls white blood cells to heal a wound or to get rid of infection. Chronic inflammation, however, can make the body more susceptible to autoimmune diseases and cancer, such as causing inflammatory bowel disease which if uncontrolled can cause colon cancer.

Harmful effects of Bacteria

Food Poisoning

Of course, all activities of bacteria are not beneficial. Some saprophytic bacteria cause decay of our food and make it unpalatable. The activities of certain bacteria produce powerful toxins such as ptomains in the food.

These toxins are powerful enough to cause food poisoning which results in serious illness and even death. Some species of Staphylococcus are the common offenders.

There is another dangerous food poisoning bacterium known as Clostridium botulinium. It causes botulism—a fatal form of food-poisoning.

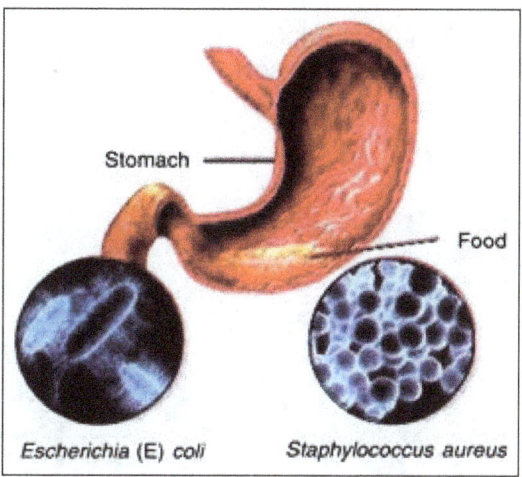

Process of food poisoning.

Disease

Many parasitic bacteria are the causative agents of bacterial diseases. They cause diseases of our economic plants, domesticated animals and man. T.J. Burrill in 1878 first gave the information that bacteria cause plant diseases.

There are more than 170 species of bacteria which cause plant diseases. Usually they are rod-like and non-spore forming. Many of them have flagella.

Fire Blight of Apple.

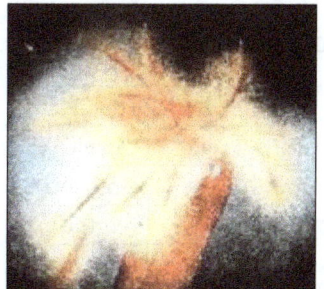

Hairy Root.

The bacteria gain entry into the host through wounds or natural openings such as stomata, lenticels, hydathodes or through the thin epidermis.

The bacterial diseases of plants belong to the following categories:

- Wilt diseases caused by blocking of the vessels of host plant by masses of bacteria. The common example of this category are the wilt diseases of potato, cucumber, water melon and eggplant.

- Crown gall and Hairy root diseases. These are due to overgrowth or hyperplasia. The crown gall of beets and hairy root of apple are the examples.

- Narcotic blights, leaf spots and rots caused by killing of parenchyma cells. Fire blight of apple, and pear and soft rot of carrot and turnip are the common examples. The following table gives a list of some important disease-causing bacteria, host plants and diseases.

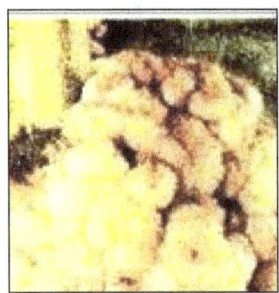

Crown Gall.

Bacterial Wilt of Egyplant.

Bacterial Wilt of Potato.

Table: Field Photographs of Bacterial Diseases of Plants.

Bacteria	Host Plant	Disease
Xanthomonas citri	Citrus plants (all species)	Citrus canker
X. malvacearum	Gossypium	Angular leaf spot or
	(cotton)	Blackarm of cotton (Soft rot)
Psedomonas	Solanum tuberosum	Bacterial brown rot or
Solanacearum	(Potato)	Wilt disease
Eriwinia amylovora	Apple, pear and other	Fire blight
	member of Rose family	
Eriwinia carotovora	Carrots and Turnips	Soft rot

A list of principal bacterial diseases of man and animals have been given in the table given below. The most interesting thing is that these diseases are cured by the use of antibiotics. These antibiotics are prepared from the bacteria themselves which cause diseases.

Table: List of diseases in Man and Animals caused by bacteria.

Name of the disease	Causal organism
Abscesses	Staphylococcus
Anthrax	Bacillus anthracis
Bacillary dysentery	Shigella sp.
Cholera	Vibrio cholera
Diphtheria	Corynebacterium diphtheria
Enteritis in intestine	Escherichia coli
Enteric fever	Salmonella sp.
Gonorrhea	Neisseria gonorrphoear
Influenza	Haemophilus influenza
Meningitis	Neisseria meningitides
Pnemonia	Diplococcus pneumoniae
Plague	Pasteurella pestis
Rat bite fever	Pasteurell pestis
Tetanus	Clostridium tetani
T.B.	Mycobacterium tuberculosae
Undulant fever	Brucella sp.
Urinogenital infections	Klebsiella sp.
Whooping cough	Bordetella sp.
Wound infections	Proteus sp.

Photographs of Symptoms of Human Diseases caused by Bacteria

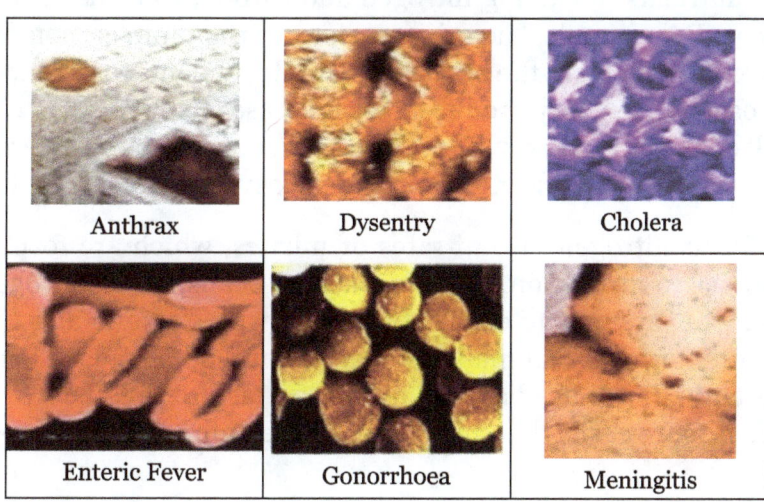

| Anthrax | Dysentry | Cholera |
| Enteric Fever | Gonorrhoea | Meningitis |

Denitrification

There are, sometimes, a group of bacteria in the soil which reverse the nitrifying process. They injure the soil by causing the loss of a part of its combined nitrogen.

This they do by breaking down nitrates into nitrites and nitrites into ammonia compounds or to free nitrogen.

$$NO_3 \rightarrow NO_2 \rightarrow NH_3 \rightarrow N \text{ gas} \uparrow$$

The free nitrogen passes into the atmosphere and is lost to the soil. The result is the lowering of soil fertility. This process is called denitrification. The bacteria which bring about denitrification are called the denitrifying bacteria.

They are most active in the soil containing excess of nitrogen compounds such as the heavily manured soils. Soils deficient in oxygen are also favourable for the activity of this type of bacteria. Denitrification is checked if the soil is well aerated by ploughing or digging and well drained. It is uneconomic to use natural and artificial nitrate manures simultaneously. The denitrifying bacteria found in the faeces contained in the manures tend to destroy the nitrates.

Bacteria and Healthy Ecosystems

Bacteria helps to maintain the health of ecosystems by breaking down dead matter and cycling nutrients into usable forms. Bacteria live in all environments, and provide most of the oxygen on Earth.

Bacteria are decomposers that break down organic materials in the soil. For instance, bacteria help to decompose dead matter, such as fallen trees, into usable nutrients for living organisms within the ecosystem. The result of decomposing dead matter results in richer soil that is able to support diverse ecosystems.

Bacteria cycles nutrients, including nitrogen and carbon, to transform the nutrients into usable forms. For example, bacteria transform carbon into carbon dioxide, which is necessary for photosynthesis, the process which allows plants to produce energy from light. Without bacteria, primary producers would cease to survive. Oxygen production is a direct result of bacterial activity. For instance, cyanobacteria, a bacteria that is present in oceans, produces a significant amount of the oxygen available on Earth.

Bacteria transforms nitrogen into nitrates or nitrites, which are forms of nitrogen that plants can use to grow. Some plants, such as beans, store nitrogen-producing bacteria in their root systems to ensure proper growth. Bacteria also live inside animals, and are used by those animals to break down food into usable nutrients. For example, cows and giraffes use bacteria in the stomach to break down plant matter so that the food is more easily digested, allowing the animals to eat low-quality foods to survive.

Role of Bacteria in Soil

Bacteria perform many important ecosystem services in the soil including improved soil structure and soil aggregation, recycling of soil nutrients, and water recycling. Soil bacteria form microaggregates in the soil by binding soil particles together with their secretions. These microaggregates are like the building blocks for improving soil structure. Improved soil structure increases water infiltration and increases water holding capacity of the soil.

Bacteria perform important functions in the soil, decomposing organic residues from enzymes released into the soil. Ingham describes the four major soil bacteria functional groups as decomposers, mutualists, pathogens and lithotrophs. Each functional bacteria group plays a role in recycling soil nutrients.

The decomposers consume the easy-to-digest carbon compounds and simple sugars and tie up soluble nutrients like nitrogen in their cell membranes. Bacteria dominate in tilled soils but they are only 20-30 percent efficient at recycling carbon (C). Bacteria are higher in nitrogen (N) content (10-30 percent nitrogen, 3 to 10 C:N ratio) than most microbes.

Of the mutualistic bacteria, there are four bacteria types that convert atmospheric nitrogen (N_2) into nitrogen for plants. There are three types of soil bacteria that fix nitrogen without a plant host and live freely in the soil and these include Azotobacter, Azospirillum and Clostridium.

Nitrogen fixing Rhizobium bacteria form nodules on a soybean root.

The *Rhizobium* bacteria (gram negative rod-shaped bacteria) species associate with a plant host: legume (alfalfa, soybeans) or clover (red, sweet, white, crimson) to form nitrogen nodules to fix nitrogen for plant growth. The plant supplies the carbon to the *Rhizobium* in the form of simple sugars. *Rhizobium* bacteria take nitrogen from the atmosphere and convert it to a form the plant can use. For plant use, the atmospheric nitrogen (N_2) or reactive nitrogen combines with oxygen to form nitrate (NO_3^-) or nitrite (NO_2^-) or combines with hydrogen to produce ammonia (NH_3^+) or ammonium (NH_4^+) which are used by plant cells to make amino acids and proteins. Figure shows nitrogen fixing bacteria.

Many soil bacteria process nitrogen in organic substrates, but only nitrogen fixing bacteria can process the nitrogen in the atmosphere into a form (fixed nitrogen) that plants can use. Nitrogen fixation occurs because these specific bacteria produce the nitrogenase enzyme. Nitrogen fixing bacteria are generally widely available in most soil types (both free living soil species and bacteria species dependent on a plant host). Free living species generally only comprise a very small percentage of the total microbial population and are often bacteria strains with low nitrogen fixing ability.

Nitrification is a process where nitrifying bacteria convert ammonia (NH_4^+) to nitrite (NO_2^-) and then to nitrate (NO_3^-). Bacteria and fungi are typically consumed by protozoa and nematodes and the microbial wastes they excrete is ammonia (NH_4^+) which is plant available nitrogen. Nitrite bacteria (*Nitrosomonas*spp.) convert the ammonia into nitrites (NO_2^-) and nitrate bacteria (*Nitrobacter* spp.) may then convert the nitrites (NO_2^-) to nitrates (NO_3^-). Nitrifying bacteria prefer alkaline soil conditions or a pH above 7. Both nitrate and ammonia are plant available forms of nitrogen; however, most plants prefer ammonia because the nitrate has to be converted to ammonia in the plant cell in order to form amino acids.

Denitrifying bacteria allow nitrate (NO_3^-) to be converted to nitrous oxide (N_2O) or dinitrogen (N_2) (atmospheric nitrogen). For denitrification to occur, a lack of oxygen or anaerobic conditions must occur to allow the bacteria to cleave off the oxygen. These conditions are common in ponded or saturated fields, compacted fields, or deep inside the microaggregates of soil where oxygen is limited. Denitrifying bacteria decrease the nitrogen fertility of soils by allowing the nitrogen to escape back into the atmosphere. On a saturated clay soil, as much as 40 to 60 percent of the soil nitrogen may be lost by denitrification to the atmosphere.

Pathogenic bacteria cause diseases in plants and a good example are bacteria blights. Healthy and diverse soil bacteria populations produce antibiotics that protect the plants from disease causing organisms and plant pathogens. Diverse bacteria populations compete for the same soil nutrients and water and tend to act as a check and balance system by reducing the disease-causing organism populations. With high microbial diversity, soils have more nonpathogenic bacteria competing with the pathogenic bacteria for nutrients and habitat. *Streptomycetes* (actinomycetes) produce more than 50 different antibiotics to protect plants from pathogenic bacteria.

Lithotrophs (chemoautotrophs) get their energy from compounds other than carbon (like nitrogen or sulfur) and include species important in nitrogen and sulfur recycling. Under well-aerated conditions, sulfur-oxidizing bacteria make the sulfur more plant available while under saturated (anaerobic, low oxygen) soil conditions, sulfur reducing bacteria make sulfur less plant available.

Actinomycetes have large filaments or hyphae and act similar to fungus in processing soil organic residues which are hard to decompose (chitin, lignin, etc.). When farmers plow or till the soil, actinomycetes release "geosmin" as they die which gives fresh-ly turned soil its characteristic smell. Actinomycetes decompose many substances but are more active at high soil pH levels. Actinomycetes are important in forming stable humus, which enhances soil structure, improves nutrient storage, and increases water retention.

Soil Benefits from Bacteria

Bacteria grow in many different microenvironments and specific niches in the soil. Bac-teria populations expand rapidly and the bacteria are more competitive when easily digestible simple sugars are readily available around in the rhizosphere. Root exudates, dead plant debris, simple sugars, and complex polysaccharides are abundant is this region. About 10 to 30 percent of the soil microorganisms in the rhizosphere are acti-nomycetes, depending on environmental conditions.

Many bacteria produce a layer of polysaccharides or glycoproteins that coats the sur-face of soil particles. These substances play an important role in cementing sand, silt and clay soil particles into stable microaggregates that improve soil structure. Bacteria live around the edges of soil mineral particles, especially clay and associated organic residues. Bacteria are important in producing polysaccharides that cement sand, silt and clay particles together to form microaggregates and improve soil struc-ture. Bacteria do not move very far in the soil, so most movement is associated with water, growing roots or hitching a ride with other soil fauna like earthworms, ants, spiders, etc.

In general, most soil bacteria do better in neutral pH soils that are well oxygenated. Bacteria provide large quantities of nitrogen to plants and nitrogen is often lacking in the soil. Many bacteria secrete enzymes in the soil to makes phosphorus more soluble and plant available. In general, bacteria tend to dominate fungi in tilled or disrupted soils because the fungi prefer more acidic environments without soil disturbance. Bac-teria also dominate in flooded fields because most fungi do not survive without oxygen. Bacteria can survive in dry or flooded conditions due to their small size, high numbers, and their ability to live in small microsites within the soil where environmental con-ditions may be favorable. Once the environmental conditions around these microsites become more favorable, the survivors quickly expand their populations. Protozoa tend to be the biggest predators of bacteria in tilled soils.

In order for bacteria to survive in the soil, they must adapt to many microenvironments. In the soil, oxygen concentrations vary widely from one microsite to another. Large pore spaces filled with air provide high levels of oxygen, which favors aerobic conditions, while a few millimeters away, smaller micropores may be anaerobic or lack oxygen. This diversity in soil microenvironments allows bacteria to thrive under various soil moisture and oxygen levels, because even after a flood (saturated soil, lack of oxygen) or soil tillage (infusion of oxygen) small microenvironments exist where different types of bacteria and microorganisms may live to repopulate the soil when environmental conditions improve.

Natural succession happens in a number of plant environments including in the soil. Bacteria improve the soil so that new plants can become established. Without bacteria, new plant populations and communities struggle to survive or even exist. Bacteria change the soil environment so that certain plant species can exist and proliferate. Where new soil is forming, certain photosynthetic bacteria start to colonize the soil, recycling nitrogen, carbon, phosphorus, and other soil nutrients to produce the first organic matter. A soil that is dominated by bacteria usually is tilled or disrupted and has higher soil pH and nitrogen available as nitrate, which is the perfect environment for low successional plants called weeds.

As the soil is disturbed less and plant diversity increases, the soil food web becomes more balanced and diverse, making soil nutrients more available in an environment better suited to higher plants. Diverse microbial populations with fungus, protozoa and nematodes keep nutrients recycling and keep disease-causing organisms in check.

UNCULTURABLE BACTERIA

The bacteria that can be grown in the laboratory are only a small fraction of the total diversity that exists in nature. At all levels of bacterial phylogeny, uncultured clades that do not grow on standard media are playing critical roles in cycling carbon, nitrogen, and other elements, synthesizing novel natural products, and impacting the surrounding organisms and environment. While molecular techniques, such as metagenomic sequencing, can provide some information independent of our ability to culture these organisms, it is essentially impossible to learn new gene and pathway functions from pure sequence data. A true understanding of the physiology of these bacteria and their roles in ecology, host health, and natural product production requires their cultivation in the laboratory. Recent advances in growing these species include coculture with other bacteria, recreating the environment in the laboratory, and combining these approaches with microcultivation technology to increase throughput and access rare species. These studies are unraveling the molecular mechanisms of unculturability and are identifying growth factors that promote the growth of previously unculturable organisms.

How do we Know there are Unculturable Bacteria?

The first evidence that not all bacteria from a given environment will grow on laboratory media came from microscopy; the number of cells that were observed microscopically far outweighed the number of colonies that grow on a petri plate. Given the name "The Great Plate Count Anomaly," the magnitude of the anomaly varied by environment but could reach several orders of magnitude. While stimulating culturing efforts, this observation also raised the question of the phylogenetic identity of these bacteria that do not grow in the laboratory. It was proposed that these are dead cells and therefore would never grow, potentially explaining the anomaly without introducing novel taxa of unculturable bacteria. In fact, many of these cells were shown to be metabolically active, even though they could not replicate on laboratory media. Additional evidence for the presence of bacterial taxa that cannot be grown in the laboratory came from molecular tools. The ability to obtain DNA sequence information from an environmental sample (by PCR amplification followed by cloning or direct sequencing) allowed characterization of phylogenetically relevant markers, such as 16S rRNA gene sequences, regardless of the viability of the organism that harbored the DNA. Such analyses revealed a hidden ocean of diversity that had never been seen by cultivation. Starting from 11 bacterial phyla (the highest-level division within the bacterial kingdom) described by Woese in 1987, the number of divisions of bacteria has grown to at least 85, the majority of which have no cultured representatives. Given that these diverse groups must be growing somewhere in the environment in order for their DNA to be present to be sequenced, the point was driven home that the culturing efforts of the last 2 centuries had managed to replicate permissive growth conditions for only a small subset of the total bacterial diversity. While DNA sequencing from mixed populations is known to be subject to artifacts that can inflate the apparent diversity, careful controls have minimized this phenomenon.

Furthermore, the repeated appearance of members of the missing phyla indicates a very real presence in nature. For example, the candidate phylum TM7 has been found repeatedly in many different environments. A sequence corresponding to the 16S rRNA gene of TM7 was first found in peat bogs, and it has since been reported to be present in a multitude of diverse environments, including soil, water, waste treatment sludge, marine sponges, the human microbiome, and many others. TM7 is just one broadly distributed phylum that has resisted substantial cultivation efforts; as indicated above, most bacterial taxa have never been studied in the lab, representing enormous genetic and biochemical diversity.

What is the Significance of these Unculturable Bacteria?

One way of measuring biological diversity is counting the number of validly described species that a given branch of the tree of life possesses. For example, it is estimated that between 800,000 and 1.2 million insect species have been described to date. This large number is partly due to the attention they have received from investigators, but it is also due to the ease of access to the subject. Most insects require no or modest magnification to observe, and they are ubiquitous. However, it is also partly due to the high level of diversity that exists in nature, as insects occupy myriad niches in the ecosphere. Expected bacterial diversity is at least as high, and most likely orders of magnitude higher, as one might reasonably expect that most of those insect species harbor at least one bacterial endosymbiont that is unique to that insect in addition to the other, multitudinous niches bacteria inhabit in the environment. Given this potential diversity, it may come as a surprise to learn that there are just over 7,000 validly described bacterial species. As with the example of insects, two factors contribute to the number of named bacteria. The first is the level of effort involved in characterizing new species, and the second is the difficulty in culturing many of them (which is essential to describing bacterial species). This "high bar" of domestic cultivation is essentially unique to microbes; it is certain that the majority of validly named macroscopic organisms were described in the wild and never cultivated in the laboratory. This distinction, requiring growth in pure culture to describe microbial species, is a direct consequence of the difficulty of determining physiologically relevant information about bacteria in the absence of a pure culture. The net result is that only a tiny fraction of the total bacterial diversity has been cultured, let alone described as a species. The missing bacteria, represented both by the phyla described above and by smaller phylogenetic subdivisions, probably harbor the majority of the metabolic diversity among not only the bacteria but also among all domains of life.

Natural products from bacteria (and their derivatives) account for half of all commercially available pharmaceuticals, and it has been estimated that the total diversity of natural small molecules (under 1 kDa) from bacteria is in excess of 109 unique compounds. Consequently, one fundamental result of having limited bacterial diversity available for study and characterization has been the collapse of the antibiotic discovery pipeline. The last new class of antibiotics that was successfully developed into a clinical therapeutic was discovered in 1987; since then, only derivatives and variations of previously discovered classes have been brought to market. This has been called the "discovery void," and it is still ongoing. One important reason for this is the repeated isolation of the same culturable bacteria, producing a limited diversity of natural products. It is estimated that at the current rate of culturing novel potential antibiotic-producing bacteria, more than 107 isolates will have to be screened to find the next new class of molecules. In addition, the screening of synthetic compound libraries has not been productive beyond modifying natural product core structures. The key to overcoming these imposing challenges is changing the rate at which novel strains are isolated; in order to do this, microbiologists will need to gain access to the uncultured majority. The potential natural products from bacteria are not limited to antibiotics, however. Microbial secondary metabolites are used in organ transplantation, cancer treatment, and cholesterol control, as well as serving as insecticides, fungicides, and antiparasitics. Almost every aspect of human health would benefit from a greater diversity and availability of microbial natural products.

Approaches to Culturing the Missing Bacterial Diversity

Simulated environments. It is difficult to replicate a natural environment at an arbitrarily high level of fidelity if it is not known which of the parameters are important for the growth of a given bacterial taxon from that setting. Therefore, one alternative is to take the bacteria back to the environment to grow them, often by moving a portion of the environment into the laboratory. The challenge with this approach is to separate the isolates of interest from the general microbial population of the environment.

Bringing the environment into the laboratory. Image of a 76-l (20-gallon) aquarium for the incubation of marine sand biofilm bacteria in a simulated environment.

The aquarium contains natural seawater, beach sand, flora, and invertebrate fauna from Canoe Beach, Nahant, MA. Operated in the laboratory of Kim Lewis at Northeastern University, the simulated environment is maintained with a filter system, a protein skimmer, a circulator, and regular exchanges of water freshly collected from the beach. This system can be used as an incubation environment, and the sand sediment can be used as a source of environmental bacteria.

The groups of Epstein and Lewis designed a diffusion chamber to accomplish this by enclosing the bacteria within a semipermeable chamber such that the cells are unable to pass through the membrane barrier but nutrients and growth factors from the environment are able to enter. Before these chambers were sealed, they were inoculated with dilute suspensions of cells from marine sediment and incubated in an aquarium of seawater on a bed of sand. Microscopic examination of the chambers revealed microcolonies of bacteria growing within them, the majority of which could be further isolated and propagated by reinoculation into fresh chambers. Comparison of the number of growing microcolonies with microscopic counts of cells in the initial inoculum yielded recovery rates of up to 40%. When the same inoculum was assayed for colony production on standard petri plates, however, the recovery rate was 0.05%, consistent with earlier reports. Several of the microcolonies from the chambers were isolated in pure culture and determined to be previously uncultured bacteria. These results demonstrated that the environment itself could be a powerful tool to gain access to bacteria that could not be cultured in the lab, while a follow-up study established that the increased recovery seen in the chambers enhanced not only the total number of growing isolates but the overall diversity as well. This technique was also subsequently shown to be effective outside the marine environment. Subsurface soil was incubated in the chambers and on petri plates, resulting in novel isolates from the chambers but not the plates. The chamber was further modified into a trap, such that one of the membranes enclosing it has a pore size just large enough that filamentous bacteria can grow into the chamber by hyphal growth but nonfilamentous bacteria cannot pass through. In a comparison by 16S rRNA gene sequencing to standard actinobacterial petri plate isolation techniques performed on the same samples, the traps gave nearly complete enrichment for filamentous act-inobacteria, with greater diversity and number, including isolates of rare groups. In a similar study, Ferrari and coworkers had cultured microcolonies of rare soil bacteria by growing them on filters suspended on soil slurry in the absence of added nutrients. Micromanipulation was used to grow one isolate in pure culture, and a follow-up study showed that these microcolonies could be reliably micromanipulated for downstream cultivation.

Bringing the environment into the laboratory also served as the basis for one of the most heralded success stories of culturing a bacterium that had previously eluded microbiologists. The SAR11 clade of Alphaproteobacteria is a widespread group of free-living bacteria found predominantly in seawater but also reported to be present in freshwater lakes. Originally identified as a prime example of a ubiquitous but un-cultured bacterium of importance to primary production in the ocean, this clade is one of the most abundant proteorhodopsin-containing organisms in seawater. Proteorho-dopsin is a light-driven proton pump that was identified in the DNA sequence of an uncultured bacterium and has been proposed to have a global impact on the carbon and energy balance in the ocean. In order to cultivate this organism, Giovannoni and his group used the natural environment in the form of seawater as a growth medium, and rather than attempt to contain specific cells separate from other isolates in the en-vironmental sample, the group utilized the fact that SAR11 was one of the most abun-dant organisms in their samples. By diluting the samples "to extinction" (a process whereby a dilution sequence is carried out until only one or a few bacteria are present in a given culture volume), they were able to separate the bacteria of this clade into the wells of microtiter plates and grow them in pure culture. Provisionally named "Candi-datus Pelagibacter ubique," its cultivation allowed characterization of this clade and its role in the marine environment. Since isolation, the genome sequence has been determined, leading to insights into its nutritional requirements and improved cul-ture conditions. In addition, the role of proteorhodopsin is starting to become clear, which may lead to a better understanding of the carbon flux in the ocean. This culture technique has now led to the culture of new members of this clade from different marine regions as well as other previously uncultured isolates from both marine and freshwater environments.

A third method for separating uncultured bacteria of interest from the rest of the organisms in their environment while culturing them in a version of their natural en-vironment (imported into the lab) was also described in 2002. The group of Keller de-veloped a method for encapsulating single bacterial cells in microdroplets of solidified agarose. A dilution to extinction series was employed to limit the number of bacteria in each gel microdroplet (GMD), and after encapsulation, the GMDs were contained in a flow column bounded by membranes that retained the encapsulated bacteria. For the marine samples cultured, seawater was constantly flushed through the columns as a growth medium. After incubation, the GMDs were passed through a flow cytometer, and those that contained a microcolony were sorted into the wells of a microtiter plate, in which phylogenetic analysis and further cultivation were carried out. Interestingly, unamended seawater as a medium yielded a higher diversity than seawater with added nutrients, as shown by 16S rRNA gene sequence analysis, reinforcing the emerging view that the natural environment can be essential for growing unculturable bacteria in the laboratory. In expanded culturing efforts using solely seawater as the growth medium, very rare bacteria were isolated as microcolonies in the GMDs, including lineages of Planctomycetales that were previously uncultured, some of which were <84% related to any published 16S sequence, cultured or otherwise. This included the sequencing performed on the original seawater samples, possibly indicating that the culturing method was enriching for very rare isolates. In a related approach, Kushmaro and coworkers used agarose spheres encapsulated in polysulfone to incubate bacteria from coral mucus in their native environment—in this case, on the surface of live coral. While this technique was not compared with standard cultures and the bacteria were not subsequently cultured outside the spheres, sequence analysis of a clone library of 16S rRNA genes suggested that about half of the isolates had not been cultured before.

One further interesting approach to growing uncultured bacteria uses the bacteria themselves to determine the particular aspect of the environment that is important to their growth rather than adding the entire environment to the medium. Graf and coworkers used high-throughput sequencing of RNA transcripts (RNA-seq) to determine that an uncultured Rikenella-like bacterium in the leech gut was utilizing mucin as a carbon and energy source. Using this insight, they were able to culture this isolate on medium containing mucin. The RNA sequence information was more useful in this regard than the genomic DNA sequence, as RNA-seq indicated what genes were actually being expressed in the growing bacterium.

Coculture: These successes at bringing the environment into the laboratory demonstrated that there are critical differences between the standard laboratory media that were traditionally being used and the natural environment of unculturable bacteria. But what were these differences? If they could be identified, it would convey a molecular understanding of unculturability and allow the synthesis of new media that did not depend on being able to reproduce the environment in the laboratory. Observations were made in the course of environmental culturing efforts that led to the identification of some of these missing factors and their source.

It was noted that some bacteria isolated from the chambers developed by Lewis and Epstein would not grow on a petri plate unless they were growing close to other bacteria from the same environment, thereby demonstrating coculture dependence for these isolates. Similarly, the widespread photosynthetic marine bacterial genus Prochlorococcus was being cultured in both natural and synthetic seawater in the hopes that it could be used as a model for marine microbial ecology. While this organism had been cultured from the ocean many times, only two variants had ever been grown in pure culture. The other isolates were all dependent on heterotrophic bacteria for coculture, and it was nearly impossible to grow these organisms from a single cell as a colony on a petri plate. A number of subsequent efforts to culture other bacteria also revealed plentiful examples of coculture-dependent isolates. The group of Kamagata observed increased growth of the previously uncultured isolate Catellibacterium nectariphilum from a sewage treatment plant in the presence of spent medium from another bacterium, while Sung and coworkers identified a number of anaerobic thermophiles in the family Clostridiaceae that were dependent on cell extract from Geobacillus toebii. In these latter cases, the growth-promoting factors have yet to be identified. However, the early examples of apparent coculture dependence in both Prochlorococcus and marine sediment bacteria led Zinser, Epstein, and Lewis to use these systems to identify the molecular mechanisms of unculturability in these strains.

In the case of Prochlorococcus, the group of Zinser set out to separate a dependent strain of Prochlorococcus (MIT9215) from its heterotrophic helpers to determine the nature of the help provided. They employed an elegant technique whereby they selected for streptomycin-resistant mutants among the abundant Prochlorococcus cells in their coculture, while the smaller population of the helpers was not large enough to contain a spontaneous resistant mutant. They were consequently able to kill the helper population by treatment with streptomycin, resulting in an apparently pure Prochlorococcus culture. By maintaining this culture of MIT9215 at high density (minimal dilution on subculture), they were able to propagate it in pure culture. However, if the concentration of Prochlorococcus cells was diluted to below about 105 cells/ml, the culture would fail to grow. In addition, MIT9215 could not form isolated colonies on petri plates without adding back the helper bacteria. After determining that a large number of different helpers would allow the growth of a variety of Prochlorococcus strains, the researchers made the intellectual leap to test whether these helper bacteria were reducing oxidative stress and thus allowing the sensitive Prochlorococcus to grow. Adding catalase, an enzyme that breaks down hydrogen peroxide, allowed improved growth of MIT9215 on plates. In a following study, the same group demonstrated that removal of H_2O_2 is both necessary and sufficient for the helping effect. Significantly, they also showed that the natural level of H2O2 in surface seawater may be high enough that this helping effect may be required in the natural environment, indicating a possible evolutionary dependence on a functional microbial ecology. At this stage, it appears that the case of Prochlorococcus is not one of growth factor contribution from the helper but rather one of environment modification.

To place this view in context, it is necessary to note that coculture dependence had been seen before; the classic example that is used in microbiology classes is the dependence of Haemophilus influenzae on Staphylococcus aureus. In this case, H. influenzae was found to need an exogenous source of both heme and NAD (termed "cozymase" at the time), originally referred to as "X" and "V" factors before they were identified, for aerobic growt. The heme is released from blood added to the medium, and NAD is released by S. aureus. While blood normally contains NAD, blood from sheep and other animals commonly used as a source of blood for culture media also contains enzymes that can destroy this factor. The dependence of H. influenzae on S. aureus can be overcome by preheating the blood used in the petri plates (creating the oddly named "chocolate" agar); this heat treatment both releases heme and inactivates the enzyme that breaks down NAD. Perhaps because the host, rather than S. aureus, was apparently the ultimate source of these factors, these observations did not directly lead to systematic coculturing efforts for other bacteria. Other examples of coculture dependence were found to consist of complementing auxotrophies, such as missing amino acids or other metabolites that the dependent bacterium could not synthesize for itself. As these auxotrophies could generally be overcome by using sufficiently rich medium (for example, yeast extract), they also did not cause a paradigm shift in culturing efforts. In an example that appears to be a case of a helper being necessary only for laboratory culture, the isolation of coculture-dependent Symbiobacterium thermophilum from compost in 1988 by Beppu and colleagues presented a mystery as to the identity of the helping factor for almost 2 decades (80). This isolate was found to be dependent on a Bacillus thermophile, and in 2006, the original group determined that this helper was providing carbon dioxide, a likely component of its natural environment. While this growth-inducing effect may be restricted to laboratory culture, the Bacillus species helper also appears to ameliorate the effects of toxic metabolites produced by S. thermophilum, leaving open the possibility of significant interactions in the environment.

With the success of coculturing efforts described above, however, the Lewis, Epstein, and Clardy groups undertook a study to directly isolate bacteria from intertidal sand biofilm that would grow only in the presence of helper organisms from the same environment in order to identify further molecular mechanisms of unculturability. The screen was based on the hypothesis that on a "crowded" isolation plate (a petri plate in which the environmental inoculum had been spread at a concentration such that a few hundred colonies would grow), some of the colonies would be growing only because they happened to be close to a helper colony. To identify these isolates, candidate colony pairs were cross-streaked and visually screened for dependent growth of one of the bacteria. Perhaps surprisingly, given the random pairing utilized, up to 10% of the screened isolates showed dependent growth. One dependent strain, Maribacter polysiphoniae KLE1104, was chosen as a model to identify its mechanism of codependence. In addition to being helped by other bacteria from the environment, it was also able to grow near a laboratory strain of Escherichia coli on a petri plate. This allowed the use of E. coli mutants to identify the genes involved in producing the growth factor and revealed that gene knockouts in the enterobactin synthesis pathway rendered E. coli unable to induce the growth of KLE1104. Enterobactin is a siderophore, a class of secreted small molecules that are able to solubilize oxidized iron (Fe^{3+}) and thereby make this essential nutrient available to cells. Spent medium from the natural environmental helper Micrococcus luteus KLE1011 was subjected to an iron-binding assay-guided fractionation to elucidate the structure of the siderophores produced in the sand biofilm, revealing five novel structural modifications of the siderophore desferrioxamine. Further screens revealed a number of bacteria from this environment to be dependent on siderophores, with a range of specificities in the ability to use different siderophores. Importantly, the addition of reduced iron (Fe^{2+}, a form of iron that is bioavailable but essentially nonexistent in the aerobic marine environment) was able to overcome these dependencies and allowed the researchers to bypass the specificity inherent in siderophores. Using this soluble form of iron, they isolated several rare bacteria, including a member of the Verrucomicrobia, a member of the Parvularculaceae, and a bacterium distantly related to the Gammaproteobacteria.

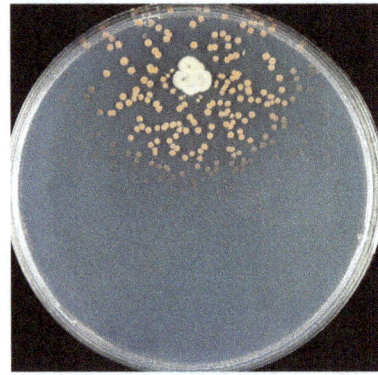

Coculture-dependent growth of an unculturable isolate. This petri plate shows a helper strain inducing the growth of a previously unculturable bacterium. A freshwater sediment isolate (a relative of Bacillus marisflavi that was originally isolated from Punderson Lake State Park, Newbury, OH) was diluted and spread evenly over the entire petri plate (R2A medium), and a culture of a helper (a relative of Bacillus megaterium that was isolated from the same environment) was spotted on the plate. The unculturable isolate grows only close to the helper, where a growth factor has diffused into the medium of the plate. Preliminary results indicate that the growth factor is a siderophore.

Helper-dependent isolates apparently have lost the ability to perform essential processes on their own, and therefore, they also have lost the ability to grow in new environments in which their helpers are not present. While it is possible to explain such a loss as evolutionary "cheating," whereby these bacteria escape the cost of performing these functions on their own by pirating them from their neighbors, it is also possible that they are actually using the presence of specific helpers as an indication (i.e., a signal) of a conducive environment to begin growth. An alternative to these hypotheses, called the "Black Queen Hypothesis" in an analogy with the card game of the same name, was recently proposed by Morris, Lenski, and Zinser. According to this theory, organisms lose functions that are being complemented by their neighbors, similar to the cheater hypothesis. However, they suggest that this dependence on helper organisms is adaptive, in that it creates specialization within microbial communities that are driven by individual selection, resulting in greater fitness for the entire community. It seems likely, given the apparent ubiquity of the phenomenon, that all of these explanations play a role under the appropriate conditions.

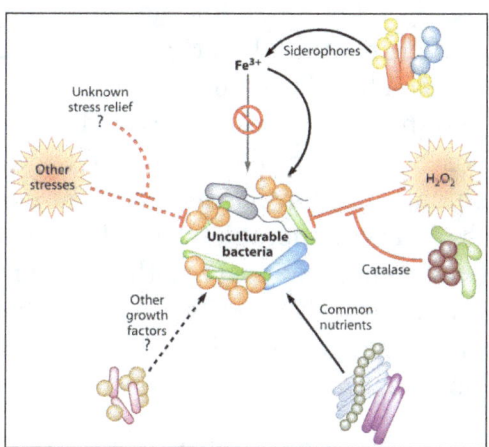

Model of mechanisms of coculture helping. Unculturable bacteria are schematically represented in the center of the figure, while known and potential helpers are arrayed around the periphery. Arrows indicate positive growth effects; stopped lines indicate inhibitory growth effects. Dashed arrows and inhibition lines indicate effects caused by as-yet-unidentified factors, while solid symbols indicate known factors. (Top center) In aerobic environments, molecular iron is completely oxidized (Fe^{3+}) and is unavailable to cells without specific systems for acquiring it (gray arrow with prohibition symbol). Siderophores from neighboring bacteria bind and solubilize Fe^{3+}, making it available to bacteria that cannot grow without this help. (Top right) Hydrogen peroxide (and possibly other forms of reactive oxygen species) can prevent the growth of sensitive bacteria. Helper organisms can protect against this effect by removing the oxidative stress, allowing the growth of the sensitive bacteria. (Bottom right) Helper bacteria can provide amino acids, vitamins, carbon sources, and other common nutrients that are often included in rich laboratory medium. (Bottom left and top left) Depictions of growth factors and stress-relieving effects yet to be discovered.

Host-associated environments. The coculture dependence and environmental incubation requirements focused on non-host-associated environments. However, as increasing focus is placed on bacteria from the human microbiome that are important in health and disease, the potential role of host-associated unculturable bacteria is being investigated. In an attempt to quantify the role of the culturable and unculturable members of the gut microbiome, the group of Gordon used germfree mice as hosts for transplanted human intestinal microbial communities. Fecal samples from donors were either directly inoculated into germfree mice or first passed through a petri plate-based culturing step before introduction into mice. Interestingly, the petri plate cultivation step resulted in a high density of colonies on each plate (about 5,000) which may have resulted in the growth of helper-dependent bacteria which otherwise would not have grown on laboratory media. The mice were assayed for both the composition of the microbiome and the weight gain of the fat pad, a measure of health. Mice colonized with bacteria that passed through the cultivation step showed equivalent fat pad weight gain to those that were directly colonized, a finding which led the authors to conclude that the uncultured component may not be critical for at least some aspects of host health. However, their results on how the composition of the microbiome responded to dietary perturbation showed that mice that received a direct transfer of bacteria had a stronger response than those who received the cultured bacteria, indicating that some functionality of the microbiome may have been lost in the cultivation step. Therefore, the questions of how much of the gut microbiome can be cultured in the laboratory and what role the remaining fraction plays in host health remain mostly unresolved. To address these issues, studies to rigorously identify the culturable fraction and determine how to grow the unculturable remainder will need to be paired with comprehensive, long-term monitoring of host health.

There is already direct evidence that coculture relationships are at work in the human microbiome. The Wade group has grown previously unculturable cluster A Synergistetes isolates from subgingival plaque in coculture with other bacteria from the mouth. In another approach to access the oral microbiome, Epstein and coworkers developed a miniature version of the trap described above that could be carried in a volunteer's mouth. The diversity obtained from the trap was compared with dilution to extinction and direct pour plating in petri plates. Both the trap and the dilution protocols produced greater diversity than the pour plates, and 10 novel isolates were cultivated. Interestingly, there was little overlap in isolates from the different culturing methods, suggesting that multiple approaches may yield greater diversity. Given the attention to the human microbiome and the culturing advances under way, it is likely that identification of growth factors contributed by one bacterial member of the microbiome to another will happen soon.

A subset of host-associated bacteria appear to be obligate intracellular symbionts or pathogens and, as such, have not been grown outside the host or cultured cell lines. A classic example is the pathogen Treponema pallidum subsp. pallidum, the causative agent of syphilis. Identified as the cause of the disease in 1905, the genome of T. pallidum subsp. pallidum was sequenced in 1998, and yet even in tissue culture with host cells, this bacterium cannot be kept in continuous culture. To date, the only reliable method of propagation is live rabbits. More-recently identified uncultured symbionts of the gut epithelium in many animals (although apparently not humans) are the segmented filamentous bacteria (SFB; also "Candidatus Arthromitus"), which appear to be epicellular on host cells. The genome sequences of mouse- and rat-associated SFB were recently completed, and like the syphilis organism, they exhibit the markedly reduced metabolic capacity characteristic of obligate, host cell-associated bacteria. SFB also have not been propagated outside host animals. The most abundant examples of bacteria closely tied to their hosts are the endosymbiotic bacteria of insects, the vast majority of which have never been cultured. The study of these numerous and biologically significant microbes has become its own field and has been reviewed recently. Examples such as these indicate, perhaps unsurprisingly, that once genome reduction occurs in a bacterium due to close host association, the difficulty of cultivation becomes significantly greater. This area may very well prove to be the greatest challenge in bacterial cultivation and therefore see the slowest progress.

Promising approaches for the future. The successes have, for the most part, combined traditional culturing methods (petri plates, liquid cultures) with new ways of making the medium more similar to the environment, including coculture with other environmental bacteria. The approach of Keller, however, in which cells were encapsulated in microdroplets, may foreshadow the next generation of culturing technology. One reason for this prediction is that a primary concern for the cultivation of novel bacteria is the effective throughput rate; as discussed above, the number of bacterial taxa that have never been cultured is thought to be considerable. Except for the cases in which cultivation efforts are directed at specific, preidentified strains of high biological importance, high-throughput methods will need to be developed to grow significant numbers of previously unculturable taxa. In general, efforts are under way to meet this need by combining the discoveries in growth factors and environment mimicry with highly parallel culturing, microfluidics, or both.

Highly parallel (macroscale) culture systems allow many isolates to be cultivated simultaneously, and there are a number of systems being developed. Tsuneda and coworkers developed a capillary-based culturing system based on porous hollow-fiber membranes. Microbial cells from an environmental sample are diluted and loaded into the fibers by syringes, and the fibers can then be lowered into a liquid environment, either simulated or natural. Dilution results in potentially single cells in many of the 96 parallel fibers, and diffusion through the porous membrane walls of the fibers allows chemical communication with the environment. After 2 months of incubation in three test environments (tidal sediment, waste treatment sludge, and a laboratory bioreactor), the fiber-based system cultivated a higher proportion of novel isolates (<97% 16S rRNA gene sequence similarity to cultured strains) than petri plates containing media designed to mimic each environment. Effectively, this created a higher-throughput system analogous to the environmental chambers. In a direct extension of the environmental chamber concept previously described, the Epstein and Lewis groups developed the isolation chip (ichip) for high-throughput cultivation. Essentially many small chambers arranged on a single substrate, the ichip contains 384 holes that form the chambers, each 1 mm in diameter. The ichip was tested on soil and seawater samples and, compared to standard petri plates or a single chamber of the original design, allowed cultivation of greater total numbers of cells as well as greater total diversity of taxa.

As the sizes of the individual cultivation chambers decrease, diffusion with the surrounding environment should increase, potentially providing an additional advantage beyond that of increased throughput. Microfluidics-based cultivation miniaturizes the cell handling and incubation of microbes, potentially maximizing both of these advantages. Microfluidic devices handle liquid down to picoliter volumes, allowing the isolation and manipulation of single cells. One class of microfluidics, droplet-based microfluidics, has been particularly promising for cultivation systems. Early work has primarily demonstrated cell handling and isolation for cultivation. The group of Köhler developed a segmented flow chip, for which aqueous "segments" (short intervals of water separated within a stream of an alkane), flow through microchannels in a silicone wafer chip. Aqueous segments could be formed holding one or a few bacterial cells by dilution of a sample and subsequently incubated to allow growth of those cells in the 20- to 60-nl volume of the segment. This allowed the parallel cultivation of potentially pure cultures in a small volume, which could then be plated after multiplication. Ismagilov and coworkers used another segment-based microfluidics approach, called a "chemistrode," in which after multiplication of bacteria within the aqueous compartments, the segments could be split to allow multiple parallel processes to be carried out on a single isolate.

This approach may allow analyses that would normally be lethal to the bacteria (such as identification of the isolate by fluorescence in situ hybridization, PCR amplification of the 16S rRNA gene for sequencing, or other techniques) to be carried out in parallel with cultivation. This group also subsequently showed (using a different microfluidics technique) that confinement of one or a few bacteria in a very small volume resulted in a high effective cell density per unit volume, which may trigger quorum-sensing activation in a single cell. It has been noted that some bacteria appear to grow only when inoculated above a certain density; this raises the possibility that confinement in small volumes by itself may allow the cultivation of previously unculturable taxa, even when only a single cell is available. Confinement in microfluidic droplets may also be used to bring together isolates for coculture. The Lin group demonstrated such a model system using auxotrophic laboratory strains in which a synthetic pair of symbiotic Escherichia coli mutants were diluted into 1-nl microdroplets. Growth occurred only when both variants were present in the same drop, showing coculture on the microfluidics scale. These are just a few examples of the many approaches to single-cell isolation, both for cultivation and for biochemical analyses. Combining droplet microfluidics as with traditional cell sorting systems may allow the isolation and cocultivation of specific unculturable bacteria of interest or the high-throughput cultivation of many novel isolates.

BACTERIOLOGY

Bacteriology is a branch of microbiology that is concerned with the study of bacteria (as well as Archaea) and related aspects. It's a field in which bacteriologists study and learn more about the various characteristics (structure, genetics, biochemistry and ecology etc) of bacteria as well as the mechanism through which they cause diseases in humans and animals.

This has allowed researchers in the field to not only get a better understanding of bacteria and their characteristics (for identification and classification purposes etc), but also how to prevent/treat/manage diseases caused by these organisms.

This field has also allowed researchers to identify some of the benefits associated with these organisms leading to their application/use in various industries.

Bacteriology played an important role in the development of the fields of molecular biology and genetics.

Some of the most recent discoveries in bacteriology include:

- Plastic pollution negatively affects oxygen-producing bacteria.

- Discovery of a luminescent compound that kills antibiotic resistant Gram-negative bacteria at the University of Sheffield and Rutherford Appleton Laboratory.

- Natural environments promote the proliferation of "good" bacteria over "bad ones".

- New bacteria species of the genus Enterobacter (E. huaxiensis and E. chuandaensis) discovered in China.

The beginnings of bacteriology paralleled the development of the microscope. The first person to see microorganisms was probably the Dutch naturalist Antonie van Leeuwenhoek, who in 1683 described some animalcules, as they were then called, in water, saliva, and other substances. These had been seen with a simple lens magnifying about 100–150 diameters. The organisms seem to correspond with some of the very large forms of bacteria as now recognized.

As late as the mid-19th century, bacteria were known only to a few experts and in a few forms as curiosities of the microscope, chiefly interesting for their minuteness and motility. Modern understanding of the forms of bacteria dates from Ferdinand Cohn's brilliant classifications, the chief results of which were published at various periods between 1853 and 1872. While Cohn and others advanced knowledge of the morphology of bacteria, other researchers, such as Louis Pasteur and Robert Koch, established the connections between bacteria and the processes of fermentation and disease, in the process discarding the theory of spontaneous generation and improving antisepsis in medical treatment.

The modern methods of bacteriological technique had their beginnings in 1870–85 with the introduction of the use of stains and by the discovery of the method of separating mixtures of organisms on plates of nutrient media solidified with gelatin or agar. Important discoveries came in 1880 and 1881, when Pasteur succeeded in immunizing animals against two diseases caused by bacteria. His research led to a study of disease prevention and the treatment of disease by vaccines and immune serums (a branch of medicine now called immunology). Other scientists recognized the importance of bacteria in agriculture and the dairy industry.

Bacteriological study subsequently developed a number of specializations, among which are agricultural, or soil, bacteriology; clinical diagnostic bacteriology; industrial bacteriology; marine bacteriology; public-health bacteriology; sanitary, or hygienic, bacteriology; and systematic bacteriology, which deals with taxonomy.

The discipline of bacteriology evolved from the need of physicians to test and apply the germ theory of disease and from economic concerns relating to the spoilage of foods and wine. The initial advances in pathogenic bacteriology were derived from the identification and characterization of bacteria associated with specific diseases. During this period, great emphasis was placed on applying Koch's postulates to test proposed cause-and-effect relationships between bacteria and specific diseases. Today, most bacterial diseases of humans and their etiologic agents have been identified, although important variants continue to evolve and sometimes emerge, e.g., Legionnaire's Disease, tuberculosis and toxic shock syndrome.

Major advances in bacteriology over the last century resulted in the development of many effective vaccines (e.g., pneumococcal polysaccharide vaccine, diphtheria toxoid, and tetanus toxoid) as well as of other vaccines (e.g., cholera, typhoid, and plague vaccines) that are less effective or have side effects. Another major advance was the discovery of antibiotics. These antimicrobial substances have not eradicated bacterial diseases, but they are powerful therapeutic tools. Their efficacy is reduced by the emergence of antibiotic resistant bacteria (now an important medical management problem) In reality, improvements in sanitation and water purification have a greater effect on the incidence of bacterial infections in a community than does the availability of antibiotics or bacterial vaccines. Nevertheless, many and serious bacterial diseases remain.

Most diseases now known to have a bacteriologic etiology have been recognized for hundreds of years. Some were described as contagious in the writings of the ancient Chinese, centuries prior to the first descriptions of bacteria by Anton van Leeuwenhoek in 1677. There remain a few diseases (such as chronic ulcerative colitis) that are thought by some investigators to be caused by bacteria but for which no pathogen has been identified. Occasionally, a previously unrecognized diseases is associated with a new group of bacteria. An example is Legionnaire's disease, an acute respiratory infection caused by the previously unrecognized genus, Legionella. Also, a newly recognized pathogen, Helicobacter, plays an important role in peptic disease. Another important example, in understanding the etiologies of venereal diseases, was the association of at least 50 percent of the cases of urethritis in male patients with Ureaplasma urealyticum or Chlamydia trachomatis.

Essentially, medical bacteriology is a branch of bacteriology that gives focus to disease-causing bacteria in human beings. It entails the detection and identification of various bacterial pathogens and the mechanism through which they cause various diseases.

To some extent, medical bacteriology also entails the study of the immune system which has made it possible to determine strategies of boosting natural immunity.

Some of the bacteria of medical significance include members of:

- Clostridium - e.g. C. pertfringens.

- Corynebacteria - e.g. C. diphteriae.

- Listeria - e.g. L. monocytogens.

- Erysipelothrix - e.g. E. rhusiopathiae.

- Neisseria - e.g. N.meningitidis and N.gonorrhea.

- Haemophilus - e.g. H. influenzae and H. hemolyticus.

- Brucella - e.g. B. abortus.

In medical bacteriology, understanding the relationship between these organisms (which are parasitic in nature in this case) and host (primary and secondary) is of great importance. It not only helps determine ways of treating diseases caused, but also finds measures that can be used to prevent or at least minimize infection rates.

In this regard, bacteriology is closely related to immunology and epidemiology. However, for the most part, this is largely with regards to bacterial infections and diseases.

Food and water bacteriology are some of the other branches closely related to medical bacteriology.

Food, which are important substances for life are also vehicles through which various bacteria can cause different types of infections and intoxications.

Some of the bacteria associated with food poisoning include:

- Salmonella

- Campylobacter

- E. coli

- C. perfringens

- Listeria

Food contamination with these organisms cause illnesses that range from minor ailments to serious incidents that can cause death. Although food bacteriology is closely related to medical bacteriology, with regards to infections and diseases, it's also an important field of study in the food industry.

It helps researchers in these fields determine ideal ways of food preservation. Such bacteria as Lactobacillus are used in some industries for food production.

Water bacteriology is also closely related to medical bacteriology given that water can also be a channel through which bacteria cause diseases.

The presence of E. coli and fecal coliform are good examples of water contamination. Here, bacteriology of water can not only help determine whether given sources of water are infected, but also provide means through which they can be treated to prevent diseases.

Various aerobic bacteria have been used for the treatment of wastewater as well as sewage. In some countries, this has been used to produce methane for energy.

References

- Bacteria, entry: newworldencyclopedia.org, Retrieved 15 June, 2019

- Bacteria-good-bad-ugly: center4research.org, Retrieved 21 April, 2019

- Economic-importance-of-bacteria-microbiology, bacteria: biologydiscussion.com, Retrieved 15 January, 2019

- Bacteria-keep-ecosystems-healthy, science: reference.com, Retrieved 7 May, 2019

- Bacteriology: microscopemaster.com, Retrieved 21 July, 2019

- Bacteriology, science: britannica.com, Retrieved 18 April 2019

- Bacteriology: microscopemaster.com, Retrieved 1 June, 2019

2

Bacterial Taxonomy

The rank-based classification of bacteria is known as bacterial taxonomy. It also plays a vital role in the nomenclature of bacteria. The topics elaborated in this chapter will help in gaining a better perspective about bacterial taxonomy as well the different phyla of bacteria which are studied within this field.

The science of classification of bacteria is called bacterial taxonomy. Bacterial taxonomy (G: taxis = arrangement or order, nomos = law or nemein = to distribute or govern), in a broader sense, consists of three separate but interrelated disciplines: classification, nomenclature, and identification.

Classification refers to the arrangements of bacteria into groups or taxa (sing, taxon) on the basis of their mutual similarity or evolutionary relatedness.

Nomenclature is the discipline concerned with the assignment of names to taxonomic groups as per published rules. Identification represents the practical side of taxonomy, which is the process of determining that a particular isolate belongs to a recognized taxon. It is to mention here that the term Bacterial systematics often is used for bacterial taxonomy.

But, systematics bears broader sense than taxonomy and is defined by many as the scientific study of organisms with the ultimate object of characterizing and arranging them in an orderly fashion. Systematics therefore encompasses disciplines such as morphology, ecology, epidemiology, biochemistry, molecular biology, and physiology of bacteria.

Importance of Bacterial Taxonomy

Bacterial taxonomy, however, is important due to following reasons:

- Bacterial taxonomy senses to be a library catalogue that helps easily access large number of books. Taxonomy therefore helps classifying and arranging bewildering diversity of bacteria into groups or taxa on the basis of their mutual similarity or evolutionary relatedness.

- The science of bacteriology is not possible without taxonomy because the latter places bacteria in meaningful, useful groups with precise names so that bacteriologists can work with them and communicate efficiently.

- Bacterial taxonomy helps bacteriologists to make predictions and frame hypotheses for further research based on knowledge of identical bacteria. For convenience, the bacteriologist can predict that a bacterium in question would be possessing similar characteristics to its relative bacterium whose characteristics are already known.

- Contribution of bacterial taxonomy in accurately identifying bacteria is of practical significance. For convenience, bacterial taxonomy contributes particularly in the area of clinical microbiology. Treatment of bacterial disease often become exceptionally difficult if the pathogen is not properly identified.

Ranks or Levels of Bacterial Taxonomy

In bacterial taxonomy, a bacterium is placed within a small but homogenous group in a rank or level. Groups of this rank or level unite creating a group of higher rank or level. In bacterial taxonomy, the most commonly used ranks or levels in their ascending order are: species, genera, families, orders, classes, phyla, and domain.

Species is the basic taxonomic group in bacterial taxonomy. Groups of species are then collected into genera (sing, genus). Groups of genera are collected into families (sing, family), families into orders, orders into classes, classes into phyla (sing, phylum), and phyla into domain (the highest rank or level). Groups of bacteria at each rank or level have names with endings or suffixes characteristic to that rank or level.

Table: Taxonomic ranks or levels in ascending order.

Rank or level	Example
Species	E. coli
Genus	Escherichia
Family	Enterobacteriaceae
Order	Enterobacteriales
Class	γ-Proteobacteria
Phylum	Proteobacteria
Domain	Bacteria

Characteristics used in Bacterial Taxonomy

Several phenotypic characteristics (e.g., morphological, physiological and metabolic, ecological) and genetic analysis have been used in bacterial (microbial) taxonomy for many years.

These characteristics are assessed and the data are used to group bacteria up to the taxonomic ladder from species to domain. Classical characteristics are quite useful in routine identification of bacteria and also provide clues for phylogenic relationships amongst them as well as with other organisms.

Morphological Characteristics

Various morphological features, e.g., cell shape, cell size, colonial morphology, arrangement of flagella, cell motility mechanism, ultra structural characteristics, staining behaviour, endospore formation, spore morphology and location, and colour are normally used to classify and identify microorganisms.

Morphological characteristics play important role in microbial classification and identification due to following reasons:

- They are easily studied and analysed especially in eukaryotic microorganisms and comparatively complex prokaryotes.

- They normally do not vary greatly with environmental changes as they are resulted in by the expression of many genes and, therefore, are usually genetically stable.

- Morphological similarity amongst microorganisms often is a good indication of phylogenetic relatedness.

Some taxonomically useful morphological characteristics and their variations are shown in Table.

Table: Some taxonomically useful morphological characteristics and their variations

Characteristics	Variations
Cell morphology	
Unicellular	Cocci, bacilli, vibrious, spirilli, spirochaetes, prosthecate, stalked, sheathed.
Multicellular	Mycelial, filamentous.
Cell arrangement	Single. Pairs. Chains. Bunces. Packets,
Staining property	
Gram staining	Gram-positive, gram-negative,
Acid fast staining	Acid-fast, non-acid fast
Flagelllation	Monotrichous, lophotrichous, amphitrichous, peritrichous, endoflagellate or non-flagellate.
Motaility	Non-motile, flagellar locomotion, gliding movement, motility due ot endoflagella.
Glycocalyx	Capsule present or absent slime layer.

Spores	Non-sporing. Endospore, expospore. Conidia. Myxo-spores.
Sporangium	Shape, location of spore.
Cell inclusions	Poly β +hydroxybutyrate, volutin, polyseeharides, sulphur droplets, parasporal protein crystals.
Ultrastructural features	Surface structures of cells – flagella, pili, fimbriae, texture of slime layer.

Physiological and Metabolic Characteristics

Some physiological and metabolic characteristics are very useful in classifying and identifying microorganisms because they are directly related to the nature and activity of microbial enzymes and transport proteins.

Some most important physiological and metabolic characteristics used in microbial taxonomy are nutritional types, cell wall components, carbon and nitrogen sources, energy metabolism, osmotic tolerance, oxygen relationships, temperature relationships, salt requirements and tolerance, secondary metabolites, storage inclusions, etc.

Table: Some taxonomically useful physiological and metabolic characteristics and their variations.

Character	Variations
Nutritional type	Photolithotrophs, chemolithotrophs, photoorganotrophs, chemoorganotrophs.
Cell wall components	Peptidoglycans, teichoic and teichuronic acids, protein, polysaccharides, pseudomurein, etc.
Carbon sources	CO_2, sugars, sugar acids alcohols, polysaccharides, organic acids.
Nitrogen sources	Molecular nitrogen, ammonium salts, nitrate, organic nitrogenous compounds.
Energy metabolism	Photosynthesis, respiration, fermentation, inorganic substrate oxidation, nitrate and sulphate oxidation.
Oxygen relationship	Aerobic, microacrophilic, facultatively anaerobic, obligaterly anaerobic.
Temperature relationship	Mesophilie, facultatively thermophilie, obligately thermophilic, hyperthermophilic, psychrophilic.

Ecological Characteristics

Ecological characteristics, i.e., the characteristics of relationship of microorganisms to their environment significantly contribute in microbial taxonomy. It is because even very closely related microorganisms may vary considerably with respect to their ecological characteristics.

For convenience, microorganisms inhabiting freshwater, marine, terrestrial, and human body environments differ from one another and from those living in different environments.

However, some of the most important ecological characteristics used in microbial taxonomy are – life cycle patterns, the nature of symbiotic relationship, pathogenic nature, and variations in the requirements for temperature, pH, oxygen, and osmotic concentrations.

Genetic Analysis

Genetic analysis is mostly used in the classification of eukaryotic microorganisms because the species is defined in these organisms in terms of sexual reproduction which occurs in them. This analysis is sometimes employed in the classification of prokaryotic microorganisms particularly those that use the processes of conjugation and transformation for gene exchange.

For example, members of genus Escherichia may conjugate with the members of genera Shigella and Salmonella but not with those of genera Enterobacter and Proteus. This shows that the members of first three genera are more closely related to one another than to Enterobacter and Proteus.

Studies of transformation with genera like Bacillus, Haemophilus Micrococcus, Rhizobium, etc. reveal that transformation takes place between different bacterial species but only rarely between genera.

This provides evidence of a close relationship between species since transformation tails to occur unless the genomes are very much similar. Bacterial plasmids that carry genes coding tor phenotypic traits undoubtedly contribute in microbial taxonomy as they occur in most genera.

Molecular Characteristics

Some recent molecular approaches such as genomic DNA GC ratios, nucleic acid hybridization, nucleic acid sequencing, ribotyping, and comparison of proteins have become increasingly important and are used routinely for determining the characteristics of microorganisms to be used in microbial taxonomy.

Genomic DNA GC Ratios (G + C Content)

Genomic DNA GC ratio (G + C) is the first, and possibly the simplest, molecular approach to be used in microbial taxonomy. The GC ratio is defined as the percentage of guanine plus cytosine in an organism's DNA.

The GC Ratio the base sequence and varies with sequence changes as follows:

$$\text{Mol \% G} + \text{C} = \frac{G+C}{G+C+A+T} \times 100$$

The GC ratio of DNA from animals and higher plants averages around 40% and ranges between 30 and 50%. Contrary to it, the GC ratio of both eukaryotic and prokaryotic microorganisms varies greatly; prokaryotic GC ratio is the most variable, ranging from around 20% to almost 80%. Despite such a wide range of variation, the GC ratio of strains within a particular species is constant.

Genomic DNA GC ratios of a wide variety of microorganisms have been determined, and knowledge of this ratio can be useful in microbial taxonomy, depending on the situation. For convenience, two microorganisms can possess identical GC ratios and yet turn out to be quite unrelated both taxonomically and phylogenetically because a variety of base sequences is possible with DNA of a single base composition.

In this case, the identical GC ratios are of no use with view point of microbial taxonomy. In contrast, if two microorganisms' GC ratio differs by greater than about 10%, they will share few DNA sequences in common and are therefore unlikely to be closely related.

GC ratio data are valuable in microbial taxonomy in at least two following ways:

- They can confirm a scheme of classification of microorganisms developed using other data. If microorganisms in the same taxon vary greatly in their GC ratios, the taxon deserves to be divided.

- GC ratio appears to be helpful in characterizing bacterial genera since the variation within a genus is usually less than 10% even though the content may vary greatly between genera.

For convenience, Staphylococcus and Micrococcus are the genera of gram-positive cocci having many features in common but differing in their GC ratio by more than 10%. The former has a GC ratio of 30-38%, whereas the latter of 64-75%.

Nucleic Acid Hybridization

Nucleic acid hybridization or genomic hybridization measures the degree of similarity between two genomes (nucleic acids) and is useful for differentiating two bacteria (microorganisms). DNA-DNA hybridization is useful to study only closely related bacteria, whereas DNA-RNA hybridization helps comparing distantly related bacteria.

DNA-DNA Hybridization

Double-stranded DNA isolated from one bacterium is dissociated into single strands at appropriate temperature which are made radioactive with ^{32}P, ^{3}H, or ^{14}C. Similarly, double-stranded DNA isolated from other bacterium is dissociated into single strands which are not made radioactive. The non-radioactive single- stranded DNA molecules are first allowed to bind to a nitrocellulose filter and unbound strands are removed by washing.

Now the filter with bound strands of DNA is incubated with the radioactive single-stranded DNA under optimal conditions of annealing. Annealing is an interesting feature of single-stranded DNA in which the strands, on cooling, tend to re-associate to form double-helix structure automatically. Annealing occurs optimally when the temperature is brought to about 25 °C below the melting temperature (Tm) in a solution of high ionic concentration.

However, during incubation the radioactive single strands of DNA hybridize with non-radioactive single strands of DNA depending on their homology in the base sequence. Then the filler is washed to remove unbound radioactive DNA molecules and the radioactivity of the hybridized radioactive DNA molecules is measured.

Following this, the amount of radioactivity in the hybridized radioactive DNA molecules is compared with the control which is taken as 100%, and this comparison gives a quantitative measure of the degree of complementarity of the two species of DNA, i.e., homology between the two DNAs. The procedure is schematically shown in figure.

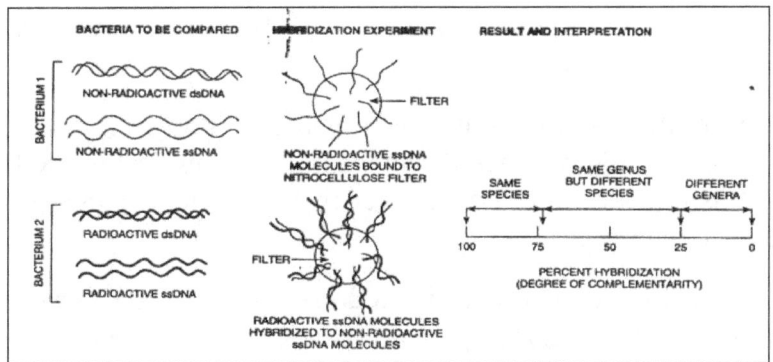

A schematic representation of DNA-DNA hybridization as a taxonomic tool.

Though there is no fixed convention as to how much hybridization between two DNAs is necessary to assign two bacteria to the same taxonomic rank, 70% or greater degree of complementarity of the two DNAs is recommended for considering the two bacteria belonging to the same species.

By contrast, degree of at least 25% is required to argue that the two bacteria should reside in the same genus. Degree of complementarity to the range of 10% or less denotes that the two bacteria are more distantly related taxonomically.

DNA-DNA hybridization is a sensitive method for revealing subtle differences in the genes of two bacteria (other microbes also) and is therefore useful for differentiating closely related bacteria.

DNA homology studies have been conducted on more than 10,000 bacteria belonging to about 2,000 species and several hundred genera. It has proved to be a powerful tool in solving many problems of bacterial taxonomy, particularly at species level.

DNA-RNA Hybridization

DNA-RNA hybridization helps compare, unlike DNA-DNA hybridization, distantly related bacteria (microorganisms) using radioactive ribosomal RNA (rRNA) or transfer RNA (tRNA).

It becomes possible because the DNA segments (genes) transcribing rRNA and tRNA represent only a small portion of the total DNA genome and have not evolved as rapidly as most other genes encoding proteins (i.e., they are more conserved in comparison to genes encoding proteins).

Among the different rRNAs, the 16S rRNA of prokaryotes and the analogous 18S rRNA of eukaryotic organisms have been found to be most suitable for comparison of their sequences is taxonomic studies. One of the major impacts of rRNA studies on taxonomy, for convenience, is the recognition of three major domains—the Archaea, the Bacteria, and the Eukarya by Woese and Colleagues in 1990.

DNA-RNA hybridization technique is similar to that employed for DNA-DNA hybridization. The filter- bound non-radioactive ssDNA is incubated with radioactive rRNA, washed, and counted.

An even more accurate measurement of complementarity is obtained by finding the temperature required to dissociate and remove half the radioactive rRNA from the filter; the higher this temperature, the stronger the DNA-rRNA complex and the more similar the base sequences.

However, DNA-RNA hybridization has been done with thousands of bacteria for relevation of their taxonomic relationships. Such studies were made with pure cultures of bacteria till 1997-98, but since then, techniques have been developed to recover rRNA genes directly from natural habitats. This has come to be called as community analysis of rRNA from natural bacterial community.

Nucleic Acid Sequencing

Nucleic acid (DNA and RNA) sequencing is another molecular characteristic that helps directly compare the genomic structures. Most attention has been given to the sequencing of 5S and 16S rRNAs isolated from the 50S and 30S subunits of 70S prokaryotic ribosome, respectively.

As mentioned in DNA-RNA hybridization, rRNAs are almost ideal for the studies of bacterial (microbial) evolution and relatedness because:

- They are essential to ribosomes found in all bacteria,

- Their functions are same in all ribosomes,

- Their structure changes very slowly with time, i.e., they are more conserved.

The procedure of rRNA sequencing involves the following steps:

- rRNA is isolated from the ribosome and purified,

- Reverse transcriptase enzyme is used to make complementary DNA (cDNA) using primers that are complementary to conserved rRNA sequences,

- The cDNA is amplified using polymerase chain reaction (PCR) and finally,

- The cDNA is sequenced and the rRNA sequence deduced from the results.

Shotgun sequencing and other genome sequencing techniques have led to the characterization of many prokaryotic genomes (approximately 100) in a very short time and many more are in the process of being sequenced. Direct comparison of complete genome sequences undoubtedly will become an important tool in determining the classificatory categories of prokaryotes.

Ribotyping

Ribotyping is a technique which measures the unique pattern that is generated when DNA from a bacterium (all other organisms also) is digested by a restriction enzyme and the fragments are separated and probed with a rRNA probe.

The technique does not involve nucleic acid sequencing. Ribotyping has proven useful for bacterial identification and has found many applications in clinical diagnostics and for the microbial analyses of food, water, and beverages.

Ribotyping is a rapid and specific method for bacterial identification; it is so specific that it has been given the nickname 'molecular fingerprinting' because a unique series of bands appears for virtually any bacterium (any organism).

In ribotyping, first the DNA is isolated from a colony or liquid culture of the bacterium to be identified. Using polymerase chain reaction (PCR), genes of DNA for rRNA (preferably 16S rRNA) and related molecules are amplified, treated with one or more restriction enzymes, separated by electrophoresis, and then probed with rRNA genes.

The pattern generated from the fragments of DNA on the gel is then digitized and a computer used to make comparison of this pattern with patterns from other bacteria available from a database.

Comparison of Proteins

The amino acid sequences of proteins are direct reflections of mRNA sequences and therefore closely related to the structures of the genes coding for their synthesis. In the light of this, the comparisons of proteins from different bacteria prove very useful taxonomically.

Although there are many methods to compare proteins, the most direct approach is to determine the amino acid sequences of proteins with the same function.

When the sequences of proteins of the same functions in two bacteria are similar, the bacteria possessing them are considered to be closely related. However, the sequences of cytochromes and other electron transport proteins, histones, heat-sock proteins, and a variety of enzymes have been used in taxonomic studies.

Classification of Bacteria on the Basis of Mode of Nutrition

Phototrops

- Those bacteria which gain energy from light.

- Phototrops are further divided into two groups on the basis of source of electron:

  ○ Photolithotrops: These bacteria gain energy from light and uses reduced inorganic compounds such as H_2S as electron source. eg. Chromatium okenii.

  ○ Photoorganotrops: These bacteria gain energy from light and uses organic compounds such as succinate as electron source.

Chemotrops

- Those bacteria gain energy from chemical compounds.

- They cannot carry out photosynthesis.

- Chemotrops are further divided into two groups on the basis of source of electron.

  ○ Chemolithotrops: They gain energy from oxidation of chemical compound and reduces inorganic compounds such as NH_3 as electron source. eg. Nitrosomonas.

  ○ Chemoorganotrops: They gain energy from chemical compounds and uses organic compound such as glucose and amino acids as source of electron. eg. Pseudomonas pseudoflava.

Autotrops

- Those bacteria which uses carbondioxide as sole source of carbon to prepare its own food.

- Autotrops are divide into two types on the basis of energy utilized to assimilate carbondioxide. ie. Photoautotrops and chemoautotrops:

    ○ Photoautotrops: They utilized light to assimilate CO_2. They are further divided into two group on the basis of electron sources. Ie. Photolithotropic autotrops and Photoorganotropic autotrops

    ○ Chemoautotrops: They utilize chemical energy for assimilation of CO_2.

Heterotrops

- Those bacteria which uses organic compound as carbon source.

- They lack the ability to fix CO_2.

- Most of the human pathogenic bacteria are heterotropic in nature.

- Some heterotrops are simple, because they have simple nutritional requirement. However there are some bacteria that require special nutrients for their growth; known as fastidious heterotrops.

Classification of Bacteria on the Basis of Optimum Temperature of Growth

Psychrophiles

- Bacteria that can grow at 0°C or below but the optimum temperature of growth is 15 °C or below and maximum temperature is 20 °C are called psychrophiles.

- Psychrophiles have polyunsaturated fattyacids in their cell membrane which gives fluid nature to the cell membrane even at lower temperature.

- Examples: Vibrio psychroerythrus, vibrio marinus, Polaromonas vaculata, Psychroflexus.

Psychrotrops (Facultative Psychrophiles)

- Those bacteria that can grow even at 0°C but optimum temperature for growth is (20-30) °C.

Mesophiles

- Those bacteria that can grow best between (25-40) °C but optimum temperature for growth is 37 °C.

- Most of the human pathogens are mesophilic in nature.

- Examples: coli, Salmonella, Klebsiella, Staphulococci.

Thermophiles

- Those bacteria that can best grow above 45 °C.

- Thermophiles capable of growing in mesophilic range are called facultative thermophiles.

- True thermophiles are called as Stenothermophiles, they are obligate thermophiles.

- Thermophils contains saturated fattyacids in their cell membrane so their cell membrane does not become too fluid even at higher temperature.

- Examples: Streptococcus thermophiles, Bacillus stearothermophilus, Thermus aquaticus.

Hyperthermophiles

- Those bacteria that have optimum temperature of growth above 80 °C.

- Mostly Archeobacteria are hyperthermophiles.

- Monolayer cell membrane of Archeobacteria is more resistant to heat and they adopt to grow in higher remperature.

- Examples: Thermodesulfobacterium, Aquifex, Pyrolobus fumari, Thermotoga.

Classification of Bacteria on the Basis of Optimum ph of Growth

Acidophiles

- Those bacteria that grow best at acidic pH.

- The cytoplasm of these bacteria are acidic in nature.

- Some acidopiles are thermophilic in nature, such bacteria are called Thermoacidophiles.

- Examples: Thiobacillus thioxidans, Thiobacillus, ferroxidans, Thermoplasma, Sulfolobus.

Alkaliphiles

- Those bacteria that grow best at alkaline pH.

- Example: vibrio cholerae: oprimum ph of growth is 8.2.

Neutrophiles

- Those bacteria that grow best at neutral pH (6.5-7.5).
- Most of the bacteria grow at neutral pH.
- Example: E. coli.

Classification of Bacteria on the Basis of Salt Requirement

Halophiles

- Those bacteria that require high concentration of NaCl for growth.
- Cell membrane of halophilic bacteria is made up of glycoprotein with high content of negatively (-Ve) charged glutamic acid and aspartic acids. So high concentration of Na^+ ion concentration is required to shield the −ve charge.
- Example: Archeobacteria, Halobacterium, Halococcus.

Halotolerant

- Most of the bacteria do not require NaCl but can tolerate low concentration of NaCl in growth media are called halotolerant.

Classification of Bacteria on the Basis of Gaseous Requirement

Obligate Aerobes

- Those bacteria that require oxygen and cannot grow in the absence of O_2.
- These bacteria carryout only oxidative type of metabolism.
- Examples; Mycobacterium, Bacillus

Facultative Anaerobes

- Those bacteria that do not require O_2 but can use it if available.
- Growth of these bacteria become batter in presence of O_2.
- These bacteria carryout both oxidative and fermentative type of metabolism.
- Examples: Coli, Klebsiella, Salmonella.

Aerotolerant Anaerobes

- Those bacteria do not require O_2 for growth but can tolerate the presence of O_2.
- Growth of these bacteria is not affected by the presence of O_2.

- These bacteria have only fermentative type of metabolism.
- Example: Lactobacillus.

Microaerophiles

- Those bacteria that do not require O_2 for growth but can tolerate low concentration of O_2.
- At atmospheric level of Oxygen growth of these bacteria is inhibited.
- These bacteria only have oxidative type of metabolism.
- Example: *Campylobacter*.

Obligate Anaerobes

- Those bacteria that can grow only in absence of Oxygen.
- Oxygen is harmful to obligate anaerobes.
- These bacteria have only fermentative type of metabolism.
- Examples: Peptococcus, Peptostreptococcus, Slostridium, methanococcus.

Capnophiles

- Those bacteria that require carbondioxide for growth.
- They are CO_2 loving organism.
- Most of the microaerophiles are capnophilic in nature.
- Example: Campylobacter, Helicobacter pylori, Brucella abortus.

Classification of Bacteria on the Basis of Morphology

Coccus

- These bacteria are spherical or oval in shape.
- On the basis of arrangement, cocci are further classified as:
  - Diplococcus: coccus in pair. eg, Neissseria gonorrhoae, Pneumococcus.
  - Streptococcus: coccus in chain. eg. Streptococcus salivarius.
  - Staphylococcus: coccus in bunch. eg. Staphylococcus aureus.
  - Tetrad: coccus in group of four.
  - Sarcina: cocus in cubical arrangement of cell. eg. Sporosarcina.

Bacilli

- These are rod shaped bacteria.
- On the basis of arrangement, bacilli are further classified as:
  - Coccobacilli: eg. Brucella.
  - Streptobacilli: chain of rod shape bacteria: eg. Bacillus subtilis.
  - Comma shaped: cg. Vibrio cholarae.
  - Chinese letter shaped: Corynebacterium dephtherae.

Mycoplasma

- They are cell wall lacking bacteria.
- Also known as PPLO (Pleuropneumonia like organism).
- Mycoplasma pneumonia.

Spirochaetes

- They are spiral shaped bacteria.
- Spirochaetes.

Rickettsiae and Chlamydiae

- They are obligate intracellular parasites resemble more closely to viruses than bacteria.

Actinomycetes

- They have filamentous or branching structure.
- They resemble more closely to Fungi than bacteria.
- Example: Streptomyces.

Classification of Bacteria on the Basis of Gram Staining

Gram positive Bacteria

- Cell wall of these bacteria is composed of peptidoglycan layer only.
- Example: Staphylococcus, Streptococcus, micrococcus.

Gram Negative bacteria

- Cell wall of these bacteria is composed of Peptidoglycan and outer membrane.
  Eg. E. coli, Salmonella.

Classification of Bacteria on the Basis of Flagella

Monotrichous Bacteria

- Bacteria having single flagella in one end of cell.
  Eg. Vibrio cholera, Pseudomonas aerogenosa.

Lophotrichous Bacteria

- Bacteria having bundle of flagella in one end of cell.
  Eg. Pseudomanas fluroscence.

Amphitrichous Bacteria

- Bacteria having single or cluster of flagella at both end of cell.
  Eg. Aquaspirillium.

Peritrichous Bacteria

- Bacteria having flagella all over the cell surface.
  Eg. E.coli, Salmonella, Klebsiella.

Atrichous Bacteria

- Bacteria without flagella.
  Eg. Shigella.

Classification of Bacteria on the Basis of Spore

Spore forming Bacteria

- Those bacteria that produce spore during unfavorable condition.
- These are further divided into two groups:

(i) Endospore Forming Bacteria:

- Spore produced within the bacterial cell.
- Bacillus, Clostridium, Sporosarcina etc.

(ii) Exospore Forming Bacteria:

- Spore produced outside the cell.

- Methylosinus.

Non Sporing Bacteria

- Those bacteria which do not produce spore.

- eg. E. coli, Salmonella.

NOMENCLATURE OF BACTERIA

Nomenclature Authority

Despite there being no official and complete classification of prokaryotes, the names (nomenclature) given to prokaryotes are regulated by the International Code of Nomenclature of Bacteria (Bacteriological Code), a book which contains general considerations, principles, rules, and various notes, and advises in a similar fashion to the nomenclature codes of other groups.

Classification Authorities

The taxa which have been correctly described are reviewed in Bergey's manual of Systematic Bacteriology, which aims to aid in the identification of species and is considered the highest authority. An online version of the taxonomic outline of bacteria and archaea is available.

List of Prokaryotic names with Standing in Nomenclature (LPSN) is an online database which currently contains over two thousand accepted names with their references, etymologies and various notes.

New Species

The International Journal of Systematic Bacteriology/International Journal of Systematic and Evolutionary Microbiology (IJSB/IJSEM) is a peer reviewed journal which acts as the official international forum for the publication of new prokaryotic taxa. If a species is published in a different peer review journal, the author can submit a request to IJSEM with the appropriate description, which if correct, the new species will be featured in the Validation List of IJSEM.

The International Code of Nomenclature of Prokaryotes (ICNP) formerly the International Code of Nomenclature of Bacteria (ICNB) or Bacteriological Code (BC) governs the scientific names for Bacteria and Archaea. It denotes the rules for naming taxa of bacteria, according to their relative rank. As such it is one of the nomenclature codes of biology.

Originally the International Code of Botanical Nomenclature dealt with bacteria, and this kept references to bacteria until these were eliminated at the 1975 IBC. An early Code for the nomenclature of bacteria was approved at the 4th International Congress for Microbiology in 1947, but was later discarded.

The latest version to be printed in book form is the 1990 Revision, but the book does not represent the current rules, as the Code has been amended since (these changes have been published in the International Journal of Systematic and Evolutionary Microbiology (IJSEM)). The 2008 Revision has since been published in the International Journal of Systematic and Evolutionary Microbiology (IJSEM). rules are maintained by the International Committee on Systematics of Prokaryotes (ICSP; formerly the International Committee on Systematic Bacteriology, ICSB).

The base-line for bacterial names is the Approved Lists with a starting point of 1980. New bacterial names are reviewed by the ICSP as being in conformity with the Rules of Nomenclature and published in the IJSEM.

Bergey's Manual of Systematic Bacteriology is the main resource for determining the identity of prokaryotic organisms, emphasizing bacterial species, using every characterizing aspect.

The manual was published subsequent to the Bergey's Manual of Determinative Bacteriology, though the latter is still published as a guide for identifying unknown bacteria. First published in 1923 by David Hendricks Bergey, it is used to classify bacteria based on their structural and functional attributes by arranging them into specific familial orders. However, this process has become more empirical in recent years.

The Taxonomic Outline of Bacteria and Archaea (TOBA) is a derived publication indexing taxon names from version two of the manual. It used to be available for free from the Bergey's manual trust website until September 2018.

Organization

The change in volume set to "Systematic Bacteriology" came in a new contract in 1980, whereupon the new style included "relationships between organisms" and had "expanded scope" overall. This new style was picked up for a four-volume set that first began publishing in 1984. The information in the volumes was separated as.

Volume 1 included information on all types of Gram-negative bacteria that were considered to have "medical and industrial importance." Volume 2 included information on all types of Gram-positive bacteria. Volume 3 deals with all of the remaining, slightly different Gram-negative bacteria, along with the Archaea. Volume 4 has information on filamentous actinomycetes and other, similar bacteria.

The current volumes differ drastically from previous volumes in that many higher taxa are not defined in terms of phenotype, but solely on 16S phylogeny, as is the case of the classes within Proteobacteria.

The current grouping is:

- Volume 1: The Archaea and the deeply branching and phototrophic Bacteria.

- Volume 2: The Proteobacteria—divided into three books:

  ∘ 2A: Introductory essays.

  ∘ 2B: The Gammaproteobacteria.

  ∘ 2C: Other classes of Proteobacteria.

- Volume 3: The Firmicutes.

- Volume 4: The Bacteroidetes, Spirochaetes, Tenericutes (Mollicutes), Acido-bacteria, Fibrobacteres, Fusobacteria, Dictyoglomi, Gemmatimonadetes, Lentisphaerae, Verrucomicrobia, Chlamydiae, and Planctomycetes.

- Volume 5 (in two parts): The Actinobacteria.

Bergey's manual of systematics of archaea and bacteria, an online book, replaces the five-volume set.

Bergey's Manual Trust

Bergey's Manual Trust was established in 1936 to sustain the publication of Bergey's Manual of Determinative Bacteriology and supplementary reference works. The Trust also recognizes individuals who have made outstanding contributions to bacterial taxonomy by presentation of the Bergey Award and Bergey Medal, jointly supported by funds from the Trust and from Springer, the publishers of the Manual.

Critical Reception

The Annals of Internal Medicine described the volumes as "clearly written, precise, and easy to read" and "particularly designed for those interested in taxonomy."

FIRMICUTES

The Firmicutes (firmus, strong, and cutis, skin, referring to the cell wall) are a phylum of bacteria, most of which have gram-positive cell wall structure. A few, however, such as Megasphaera, Pectinatus, Selenomonas and Zymophilus, have a porous pseudo-outer membrane that causes them to stain gram-negative. Scientists once classified the Firmicutes to include all gram-positive bacteria, but have recently defined them to be of a core group of related forms called the low-G+C group, in contrast to the Actinobacteria. They have round cells, called cocci (singular coccus), or rod-like forms (bacillus).

Many Firmicutes produce endospores, which are resistant to desiccation and can survive extreme conditions. They are found in various environments, and the group includes some notable pathogens. Those in one family, the heliobacteria, produce energy through anoxygenic photosynthesis. Firmicutes play an important role in beer, wine, and cider spoilage.

Classes

The group is typically divided into the Clostridia, which are anaerobic, the Bacilli, which are obligate or facultative aerobes, and the Mollicutes, which are parasites. On phylogenetic trees, the first two groups show up as paraphyletic or polyphyletic, as do their main genera, Clostridium and Bacillus.

Phylogeny

The phylogeny is based on 16S rRNA-based LTP release 132 by the All-Species Living Tree Project, with the currently accepted taxonomy based on the List of Prokaryotic names with Standing in Nomenclature (LPSN), National Center for Biotechnology Information (NCBI), and some non-validated clade names from Genome Taxonomic Database.

Genera

More than 274 genera were considered as of 2016 to be within the Firmicutes phylum notable genera of Firmicutes include:

Bacilli, Order Bacillales:

- Bacillus

- Listeria

- Staphylococcus

Bacilli, Order Lactobacillales:

- Lactobacillus

Clostridia:

- Acetobacterium

- Clostridium

- Eubacterium

- Heliobacteria

- Heliospirillum

- Megasphaera

- Pectinatus

- Selenomonas

- Zymophilus

- Sporohalobacter

- Sporomusla

Erysipelotrichia:

- Erysipelothrix

Health Implications

Firmicutes make up the largest portion of the mouse and human gut microbiome. The division Firmicutes as part of the gut flora has been shown to be involved in energy resorption, and potentially related to the development of diabetes and obesity. Within the gut of healthy human adults, the most abundant bacterium: Faecalibacterium prausnitzii (F. prausnitzii), which makes up 5% of the total gut microbiome, is a member of the Firmicutes phylum. This species is directly associated with reduced low-grade inflammation in obesity. F. prausnitzii has been found in higher levels within the guts of obese children than in non-obese children.

In multiple studies a higher abundance of Firmicutes has been found in obese individuals than in lean controls. A higher level of Lactobacillus (of the Firmicutes phylum) has been found in obese patients and in one study, obese patients put on weight loss diets showed a reduced amount of Firmicutes within their guts.

Diet changes in mice have also been shown to promote changes in Firmicutes abundance. A higher relative abundance of Firmicutes was seen in mice fed a western diet (high fat/high sugar) than in mice fed a standard low fat/high polysaccharide diet. The higher amount of Firmicutes was also linked to more adiposity and body weight within mice. Specifically, within obese mice, the class Mollicutes (within the Firmicutes phylum) was the most common. When the microbiota of obese mice with this higher Firmicutes abundance was transplanted into the guts of germ-free mice, the germ-free mice gained a significant amount of fat as compared to those transplanted with the microbiota of lean mice with lower Firmicutes abundance.

The presence of Christensenella (Firmicutes, in class Clostridia), isolated from human faeces, has been found to correlate with lower body mass index.

Laboratory Detection

The presence of Firmicutes can be reliably detected with polymerase chain reaction (PCR) techniques.

BACTERIAL PHYLA

The bacterial phyla are the major lineages, known as phyla or divisions, of the domain Bacteria.

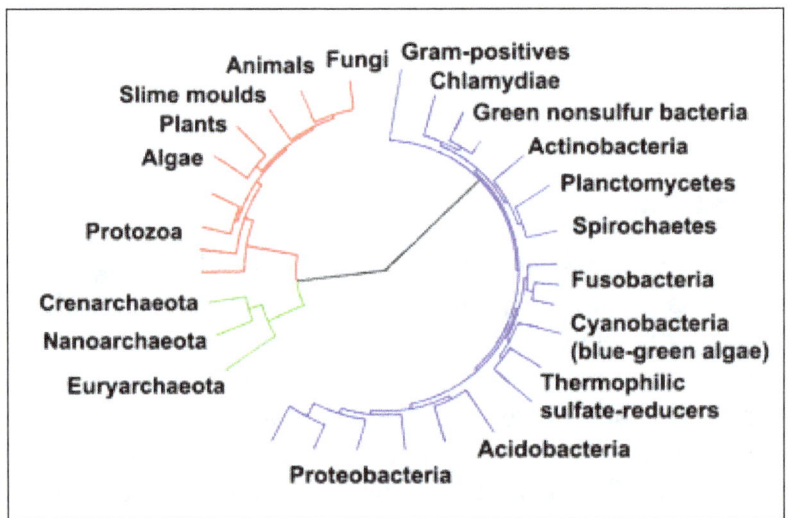

Phylogenetic tree showing the diversity of bacteria, compared to other organism
Eukaryotes are colored red, archaea green and bacteria blue.

In the scientific classification established by Carl von Linné, each bacterial strain has to be assigned to a species (binary nomenclature), which is a lower level of a hierarchy of ranks. Currently, the most accepted mega-classification system is under the three-domain system, which is based on molecular phylogeny. In that system, bacteria are members of the domain Bacteria and "phylum" is the rank below domain, since the rank "kingdom" is disused at present in bacterial taxonomy. When bacterial nomenclature was controlled under the Botanical Code, the term division was used, but now that bacterial nomenclature (with the exception of cyanobacteria) is controlled under the Bacteriological Code, the term phylum is preferred.

In this classification scheme, Bacteria is (unofficially) subdivided into 30 phyla with representatives cultured in a lab. Many major clades of bacteria that cannot currently be cultured are known solely and somewhat indirectly through metagenomics, the analysis of bulk samples from the environment. If these possible clades, candidate phyla, are included, the number of phyla is 52 or higher. Therefore, the number of major phyla has increased from 12 identifiable lineages in 1987, to 30 in 2014, or over 50 including candidate phyla. The total number has been estimated to exceed 1,000 bacterial phyla.

At the base of the clade Bacteria, close to the last universal common ancestor of all living things, some scientists believe there may be a definite branching order, whereas other scientists, such as Norman Pace, believe there was a large hard polytomy, a simultaneous multiple speciation event.

Molecular Phylogenetics

Traditionally, phylogeny was inferred and taxonomy established based on studies of morphology. Recently molecular phylogenetics has been used to allow better elucidation of the evolutionary relationship of species by analysing their DNA and protein sequences, for example their ribosomal DNA. The lack of easily accessible morphological features, such as those present in animals and plants, hampered early efforts of classification and resulted in erroneous, distorted and confused classification, an example of which, noted Carl Woese, is Pseudomonas whose etymology ironically matched its taxonomy, namely "false unit".

New Cultured Phyla

New species have been cultured since 1987, when Woese's review paper was published, that are sufficiently different that they warrant a new phylum. Most of these are thermophiles and often also chemolithoautotrophs, such as Aquificae, which oxidises hydrogen gas. Other non-thermophiles, such as Acidobacteria, a ubiquitous phylum with divergent physiologies, have been found, some of which are chemolithotrophs, such as Nitrospira (nitrile-oxidising) or Leptospirillum (Fe-oxidising)., some proposed phyla however do not appear in LPSN as they were insufficiently described or are awaiting approval or it is debated if they may belong to a pre-existing phylum. An example of this is the genus Caldithrix, consisting of C. palaeochoryensis and C. abyssi, which is considered Deferribacteres, however, it shares only 81% similarity with the other Deferribacteres (Deferribacter species and relatives) and is considered a separate phylum by Rappé and Giovannoni. Additionally the placement of the genus Geovibrio in the phylum Deferribacteres is debated.

Uncultivated Phyla and Metagenomics

With the advent of methods to analyse environmental DNA (metagenomics), the 16S rRNA of an extremely large number of undiscovered species have been found, showing that there are several whole phyla which have no known cultivable representative and that some phyla lack in culture major subdivisions as is the case for Verrucomicrobia and

Chloroflexi. The term Candidatus is used for proposed species for which the lack of information prevents it to be validated, such as where the only evidence is DNA sequence data, even if the whole genome has been sequenced. When the species are members of a whole phylum it is called a candidate division (or candidate phylum) and in 2003 there were 26 candidate phyla out of 52. A candidate phylum was defined by Hugenholtz and Pace in 1998, as a set of 16S ribosomal RNA sequences with less than 85% similarity to already-described phyla. More recently an even lower threshold of 75% was proposed. Three candidate phyla were known before 1998, prior to the 85% threshold definition by Hugenholtz and Pace:

- OS-K group (from Octopus spring, Yellowstone National Park).

- Marine Group A (from Pacific ocean).

- Termite Group 1 (from a Reticulitermes speratus termite gut, now Elusimicrobia).

Since then several other candidate phyla have been identified:

- OP1, OP3, OP5 (now Caldiserica), OP8, OP9 (now Atribacteria), OP10 (now Armatimonadetes), OP11 (obsidian pool, Yellowstone National Park).

- WS2, WS3, WS5, WS6 (Wurtsmith contaminated aquifer).

- SC3 and SC4 (from arid soil).

- vadinBE97 (now Lentisphaerae).

- NC10 (from flooded caves, paddy fields, intertidal zones, etc).

- BRC1 (from bulk soil and rice roots).

- ABY1 (from sediment).

- Guyamas1 (from hydrothermal vents).

- GN01, GN02, GN04 (from a Guerrero Negro hypersaline microbial mat).

- NKB19 (from activated sludge).

- SBR1093 (from activated sludge).

- TM6 and TM7 (Torf, Mittlere Schicht lit. "peat, middle layer").

Since then a candidate phylum called Poribacteria was discovered, living in symbiosis with sponges and extensively studied. The divergence of the major bacterial lineages predates sponges Another candidate phylum, called Tectomicrobia, was also found living in symbiosis with sponges. And Nitrospina gracilis, which had long eluded phylogenetic placement, was proposed to belong to a new phylum,

Nitrospinae. Other candidate phyla that have been the center of some studies are TM7, the genomes of organisms of which have even been sequenced (draft), WS6 and Marine Group A.

Two species of the candidate phylum OP10, which is now called Armatimonadetes, where recently cultured: Armatimonas rosea isolated from the rhizoplane of a reed in a lake in Japan and Chthonomonas calidirosea from an isolate from geothermally heated soil at Hell's Gate, Tikitere, New Zealand.

One species, Caldisericum exile, of the candidate phylum OP5 was cultured, leading to it being named Caldiserica. The candidate phylum VadinBE97 is now known as Lentisphaerae after Lentisphaera araneosa and Victivallis vadensis were cultured.

Superphyla

Despite the unclear branching order for most bacterial phyla, several groups of phyla have clear clustering and are referred to as superphyla.

FCB Group

The FCB group (now called Sphingobacteria) includes Bacteroidetes, the unplaced genus Caldithrix, Chlorobi, candidate phylum Cloacimonetes, Fibrobacteres, Gemmatimonadates, candidate phylum Ignavibacteriae, candidate phylum Latescibacteria, candidate phylum Marinimicrobia, and candidate phylum Zixibacteria.

PVC Group

The PVC group (now called Planctobacteria) includes Chlamydiae, Lentisphaerae, candidate phylum Omnitrophica, Planctomycetes, candidate phylum Poribacteria, and Verrucomicrobia.

Patescibacteria

The proposed superphylum, Patescibacteria, includes candidate phyla Gracilibacteria, Microgenomates, Parcubacteria, and Saccharibacteria and possibly Dependentiae. These same candidate phyla, along with candidate phyla Berkelbacteria, CPR2, CPR3, Kazan, Perigrinibacteria, SM2F11, WS6, and WWE3 were more recently proposed to belong to the larger CPR Group. To complicate matters, it has been suggested that several of these phyla are themselves actually superphyla.

Terrabacteria

The proposed superphylum, Terrabacteria, includes Actinobacteria, Cyanobacteria, Deinococcus–Thermus, Chloroflexi, Firmicutes, and candidate phylum OP10.

Proteobacteria as Superphylum

It has been proposed that several of the classes of the phylum Proteobacteria are phyla in their own right, which would make Proteobacteria a superphylum. It was recently proposed that the class Epsilonproteobacteria be combined with the order Desulfurellales to create the new phylum Epsilonbacteraeota.

Cryptic Superphyla

Several candidate phyla (Microgenomates, Omnitrophica, Parcubacteria, and Saccharibacteria) and several accepted phyla (Elusimicrobia, Caldiserica, and Armatimonadetes) have been suggested to actually be superphyla that were incorrectly described as phyla because rules for defining a bacterial phylum are lacking. For example, it is suggested that candidate phylum Microgenomates is actually a superphylum that encompasses 28 subordinate phyla and that phylum Elusimocrobia is actually a superphylum that encompasses 7 subordinate phyla.

Phyla

As of January 2016, there are 30 phyla in the domain "Bacteria" accepted by LPSN. There are no fixed rules to the nomenclature of bacterial phyla. It was proposed that the suffix "-bacteria" be used for phyla, but generally the name of the phylum is generally the plural of the type genus, with the exception of the Firmicutes, Cyanobacteria, and Proteobacteria, whose names do not stem from a genus name.

Acidobacteria

The Acidobacteria (diderm Gram negative) is most abundant bacterial phylum in many soils, but its members are mostly uncultured. Additionally, they are phenotypically diverse and include not only acidophiles, but also many non-acidophiles. Generally its members divide slowly, exhibit slow metabolic rates under low-nutrient conditions and can tolerate fluctuations in soil hydration.

Actinobacteria

The Actinobacteria is a phylum of monoderm Gram positive bacteria, many of which are notable secondary metabolite producers. There are only two phyla of monoderm Gram positive bacteria, the other being the Firmicutes; the actinobacteria generally have higher GC content so are sometimes called "high-CG Gram positive bacteria". Notable genera/species include Streptomyces (antibiotic production), Cutibacterium acnes (odorous skin commensal) and Propionibacterium freudenreichii (holes in Emmental).

Aquificae

The Aquificae (diderm Gram negative) contains only 14 genera (including Aquifex and Hydrogenobacter). The species are hyperthermophiles and chemolithotrophs (sulphur). According to some studies, this may be one of the most deep branching phyla.

Bacteroidetes

The Bacteroidetes (diderm Gram negative) is a member of the FBC superphylum. Some species are opportunistic pathogens, while other are the most common human gut commensal bacteria. Gained notoriety in the non-scientific community with the urban myth of a bacterial weight loss powder.

Caldiserica

This phylum was formerly known as candidate phylum OP5, Caldisericum exile is the sole representative.

Chlamydiae

The Chlamydiae (diderms, weakly Gram negative) is a phylum of the PVC superphylum. It is composed of only 6 genera of obligate intracellular pathogens with a complex life cycle. Species include Chlamydia trachomatis (chlamydia infection).

Chlorobi

Chlorobi is a member of the FBC superphylum. It contains only 7 genera of obligately anaerobic photoautotrophic bacteria, known colloquially as Green sulfur bacteria. The reaction centre for photosynthesis in Chlorobi and Chloroflexi (another photosynthetic group) is formed by a structures called the chlorosome as opposed to phycobilisomes of cyanobacteria (another photosynthetic group).

Chloroflexi

Chloroflexi, a diverse phylum including thermophiles and halorespirers, are known colloquially as Green non-sulfur bacteria.

Firmicutes

Firmicutes, Low-G+C Gram positive species most often spore-forming, in two/three classes: the class Bacilli such as the Bacillus spp. (e.g. B. anthracis, a pathogen, and B. subtilis, biotechnologically useful), lactic acid bacteria (e.g. Lactobacillus casei in yoghurt, Oenococcus oeni in malolactic fermentation, Streptococcus pyogenes, pathogen), the class Clostridia of mostly anaerobic sulphite-reducing saprophytic species, includes the genus Clostridium (e.g. the pathogens C. dificile, C. tetani, C. botulinum and the biotech C. acetobutylicum).

Proteobacteria

Proteobacteria, contains most of the "commonly known" species, such as Escherichia coli and Pseudomonas aeruginosa.

Spirochaetes

Spirochaetes, notable for compartmentalisation and species include Borrelia burgdorferi, which causes Lyme disease.

Synergistetes

The Synergistetes is a phylum whose members are diderm Gram negative, rod-shaped obligate anaerobes, some of which are human commensals.

Tenericutes

The Tenericutes includes the class Mollicutes, formerly/debatedly of the phylum Firmicutes (sister clades). Despite their monoderm Gram-positive relatives, they lack peptidoglycan.

Thermodesulfobacteria

The Thermodesulfobacteria is a phylum composed of only three genera in the same family (Thermodesulfobacteriaceae: Caldimicrobium, Thermodesulfatator and Thermodesulfobacterium). The members of the phylum are thermophilic sulphate-reducers.

Thermotogae

The Thermotogae is a phylum of whose members possess an unusual outer envelope called the toga and are mostly hyperthermophilic obligate anaerobic fermenters.

Verrucomicrobia

Verrucomicrobia is a phylum of the PVC superphylum. Like the Planctomycetes species, its members possess a compartmentalised cell plan with a condensed nucleoid and the ribosomes pirellulosome (enclosed by the intracytoplasmic membrane) and paryphoplasm compartment between the intracytoplasmic membrane and cytoplasmic membrane.

MONERA

Monera is a kingdom in biology that comprises prokaryotes, which are single-celled organism that have no true nucleus. Monera is the most ancient group of organisms on earth, as well as the most numerous. In this kingdom, the organisms have naked DNA that forms a clump called the nucleoid, as shown below, while organisms in all other kingdoms have DNA enclosed in a nucleus. Since monerans are prokaryotes, such as bacteria, they have no membrane-bound organelles. They are also microscopic and usually live in moist environments. For example, we can find monerans within bodies of animals and plants, and in hot springs. The term Monera is no longer used by many scientists, because they have found that the two groups that make up this kingdom, archaea and bacteria, aren't as closely related as once thought. Rather, archaea are closer to eukaryotes than they are to bacteria.

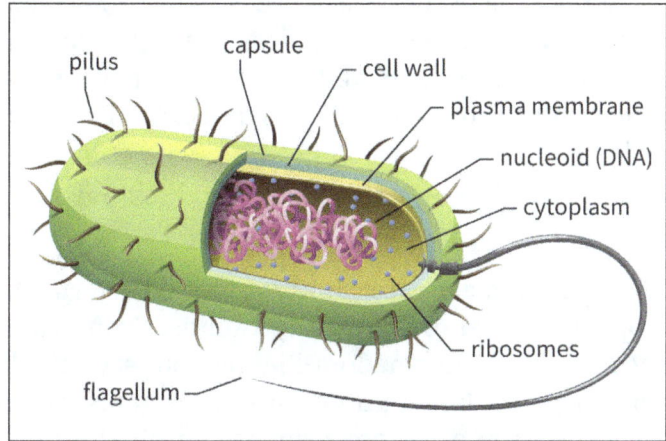

Prokaryote cell.

Characteristics of Monerans

Some monerans are autotrophic, making their own food through either chemosynthesis, like nitrifying bacteria in the nitrogen cycle, or through photosynthesis, like purple sulfur bacteria. Others are heterotrophic, either existing as saprophytic decomposers that feed on dead organic matter found in the soil, or as parasitic bacteria that acquire food from a living host, usually causing it harm in the process.

Organisms in the Monera kingdom can have different means of mobility, such as movement by using the flagella, as in the diagram above, to propel themselves through liquids, axial filaments to rotate, or by secreting slime to glide.

Most organisms in the Monera kingdom reproduce asexually through binary fission, which does not allow for much genetic diversity, since each daughter cell produced receives genetic information that is identical to the parent's.

Certain organisms in the Monera kingdom can surround themselves by a capsule as a means of defense from adverse conditions and threats, such as phagocytosis by white blood cells, and desiccation. The cell becomes coated and partially dehydrated, turning into an endospore, which a dormant phase. When conditions become favorable again, the endospore then returns to being an active cell.

Subkingdoms of Monera

Eubacteria

"Eu-" as a prefix means "true", so eubacteria is the subkingdom that includes the typical bacteria most of us are familiar with, such as E. coli.

Archeabacteria

Includes extremophiles that can tolerate harsh environments, such as high temperatures, high acidity, and the absence of oxygen. These organisms include thermophiles, or heat-loving organisms that can exist in places like volcanic vents; methanogens, which can exist in anoxic environments and produce methane gas as a metabolic by-product; and halophiles, or salt-loving organisms that can exist in places where salinity levels are incredibly high.

Cyanobacteria

Cyanobacteria, or blue-green algae, are photosynthetic bacteria. Their cells contain pigments like chlorophyll, carotenoids, and phycobilins. Cyanobacteria are aquatic in nature and were once a part of the kingdom Plantae, but scientist have since discovered that they are prokaryotes. Being photosynthetic organisms, they are important producers of oxygen, though some of them produce neurotoxins that can harm aquatic life.

How Monerans Benefit other Organisms

Bacteria enrich soil, and are very important in the nitrogen cycle, which is essential for plants survival. They are also useful to us in the way that they are important in producing some foods like cheese and vinegar, and used in the production of some antibiotics.

Methanogens also play a significant role in our lives, as they are used in the treatment of sewage. Finally, archaerbacteria as a whole group supports ecosystems in habitats with extreme conditions, since many organisms rely on them as a source of food.

References

- Bacterial-taxonomy-meaning-importance-and-levels, bacterial-taxonomy, bacteria: biologydiscussion.com, Retrieved 28 February, 2019

- Madigan, michael (2009). Brock biology of microorganisms. San francisco: pearson/benjamin cummings. Isbn 978-0-13-232460-1

- Lake, j. (2009). "evidence for an early prokaryotic endosymbiosis". Nature. 460 (7258): 967–971. Bibcode:2009natur.460..967l. Doi:10.1038/nature08183. Pmid 19693078

- Classification-of-bacteria: onlinebiologynotes.com, Retrieved 19 May, 2019

- Garrity gm, lilburn tg, cole jr, harrison sh, euzeby j, tindall bj (2007). "taxonomic outline of the bacteria and archaea, release 7.7". Michigan state university board of trustees. Doi:10.1601/toba7.7

- J. P. Euzéby. "firmicutes". List of prokaryotic names with standing in nomenclature (lpsn). Archived from the original on january 27, 2013. Retrieved 2013-03-20

- Million, m. (april 2013). "gut bacterial microbiota and obesity". Cell microbiology and infection. 19 (4): 305–313. Doi:10.1111/1469-0691.12172. Pmid 23452229

- Monera: biologydictionary.net, Retrieved 5 May, 2019

3

Gram Negative and Gram Positive Bacteria

Gram negative bacteria are the bacteria which do not retain the crystal violet stain which is used in the gram-staining method of bacterial differentiation. The bacteria that give a positive result in the Gram stain test are known as gram positive bacteria. The topics elaborated in this chapter will help in gaining a better perspective about the different types of gram negative and positive bacteria.

Gram Postive

The phrase 'gram-positive' is a term used by microbiologist to classify bacteria into two groups (gram-positive or gram-negative). This positive/negative reference is based on the bacterium's chemical and physical cell wall properties.

To determine if a bacteria is gram-positive or gram-negative a microbiologist will perform a special type of staining technique, called a Gram Stain. The name comes from its discoverer and inventor, Han Christian Gram. This stain will either stain the cells purple (for positive) or pink (for negative).

Gram-positive bacteria have a very thick cell wall made of a protein called peptidoglycan. These bacteria retain the crystal violet dye (one of the 2 main chemicals used for gram staining). Whereas, gram-negative bacteria have a very thin peptidoglycan layer that is sandwiched between an inner cell membrane and a bacterial outer membrane.

Gram Negative

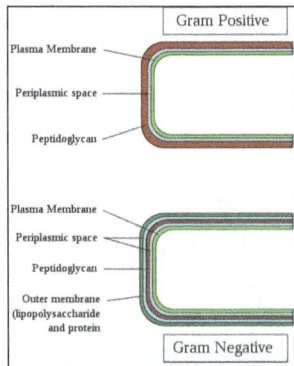

Gram-negative bacteria do not retain the crystal violet stain because of its physical makeup and will stain pink. Thus, why they are called gram-negative. A microscope to see this stain coloring. The microbes are still very tiny and are usually placed on a slide prior to staining.

Are they Good or Bad?

Usually, gram-positive bacteria are the helpful, probiotic bacteria we hear about in the news, like LAB. They are the happy ones that live in our gut and help us digests food. Gram-negative bacteria, by coincidence, are usually thought of as the nasty bugs that can make us sick and can be harmful. For example, several species of Escherichia coli (E. coli), are common causes of food-borne disease. Another example is Vibrio cholera, the bacteria responsible for cholera, is a waterborne pathogen.

However, this is just a generality and cannot be assumed that all gram-negative bacteria are harmful. Gram-positive bacteria can also be pathogenic. Clostridium botulinum, the bacterium responsible for producing neurotoxins that can kill in hours is a gram-positive bacterium.

GRAM NEGATIVE BACTERIA

Gram negative bacteria lose the crystal violet stain (and take the color of the red counterstain) in Gram's method of staining. This is characteristic of bacteria that have a cell wall composed of a thin layer of a particular substance (called peptidoglycan).

The Gram-negative bacteria include most of the bacteria normally found in the gastrointestinal tract that can be responsible for disease as well as gonococci (venereal disease) and meningococci (bacterial meningitis). The organisms responsible for cholera and bubonic plague are Gram-negative.

The Danish bacteriologist J.M.C. Gram devised this method of staining bacteria using a dye called crystal (gentian) violet. Gram's method helps distinguish between different types of bacteria. The gram-staining characteristics of bacteria are denoted as positive or negative, depending upon whether the bacteria take up and retain the crystal violet stain or not.

GRAM STAINING

Gram staining is a type of differential staining used to distinguish between Gram positive and Gram negative bacterial groups, based on inherent differences in their cell wall constituents.

Stains increase contrast allowing bacteria to be observed with greater ease. Gram staining is dependent on the interaction between the negatively charged surface of bacteria and the positive ions within common stains.

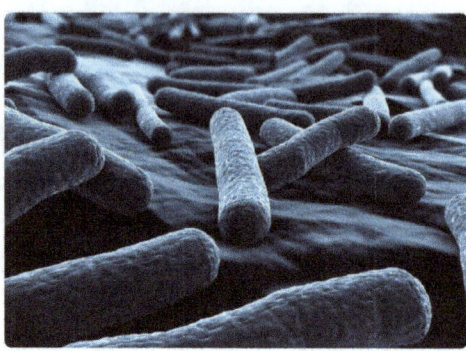

Gram negative bacteria are often pathogenic and contain endotoxins within their outer membranes. The cell wall structure of Gram negative bacteria also confers an increased ability to resist antibiotics. Escherichia coli is an example of a common Gram negative bacterium.

Gram Stain Procedure

The Gram stain procedure was developed by Hans Christian Gram in the 1800s and was one of the first techniques taught to students in microbiology laboratories. A dried bacterial smear is flooded with a crystal violet stain before iodine is added to fix the stain into an insoluble complex.

The smear is de-stained using an alcohol or acetone solution to remove any weakly bound stain, before being counterstained with the red dye, safranin. The slide is rinsed between each procedural step with the final wash acting as preparation for observation under a microscope.

Gram positive bacteria are observed as blue or violet colored cells (having retained the stain) whilst Gram negative bacteria appear pink because of the counterstain.

Gram Negative Bacteria Cell Wall Structure

Gram negative bacteria are characterized by the presence of the periplasmic space, which is a single layer of peptidoglycan sandwiched between the cytoplasmic membrane and the outer membrane. Peptidoglycan, also known as murein, is a polymer that consists of a carbohydrate backbone and amino acids.

Peptide chains within the peptidoglycan structure are partially cross-linked in Gram negative bacteria, contrasting with the highly cross-linked peptide chains of Gram positive bacteria.

The outer membrane contains lipopolysaccharide, a large molecule that is toxic to animals. During Gram staining, the outer membrane of Gram negative bacteria deteriorates from the alcohol added to the sample and the thin layer of peptidoglycan is not able to retain the crystal violet stain.

The counterstain is added to provide contrast by staining the decolorized Gram negative bacteria through the thin peptidoglycan layer whilst being light enough to not disturb the crystal violet staining on the Gram positive bacteria.

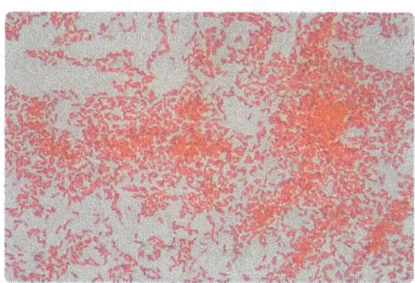

Gram Negative Bacteria as Pathogens

Gram negative bacteria are often pathogenic and include Escherichia coli, a common cause of food poisoning and Vibrio cholerae, the waterborne pathogen responsible for cholera outbreaks. The pathogenic capability of Gram negative bacteria is caused by their constituent membrane components.

The lipopolysaccharide endotoxin that resides in the outer membrane can cause a toxic reaction or strong immune response in the host animal. Gram-negative bacteria that enter the circulatory system may release lipopolysaccharides in large enough amounts to trigger an immune response that is injurious to the host's organs and tissues.

The circulating lipopolysaccarides found within the blood stream of patients suffering from sepsis indicates that endotoxins are an important drug target for the prevention and treatment of septic shock.

Gram Negative Bacteria and Resistance to Antibiotics

Gram- negative bacteria are less susceptible to antibiotics because of their outer membrane. This is because the outer membrane provides protection from treatments that would ordinarily damage the inner membrane. Antibiotic resistant strains of bacteria have been noted to display modifications in the lipid or protein composition of the outer membrane.

Small hydrophilic antibiotics target porins, protein channels that provide a pathway through the outer membrane. Antibiotic resistance has been linked to a reduction in the rate of entry for antibiotics through porins via modifications to the outer membrane profile or reduced permeability caused by specific mutations.

SALMONELLA

Salmonella is a genus of rod-shaped (bacillus) Gram-negative bacteria of the family Enterobacteriaceae. The two species of Salmonella are Salmonella enterica and Salmonella bongori. S. enterica is the type species and is further divided into six subspecies that include over 2,600 serotypes. Salmonella was named after Daniel Elmer Salmon, an American veterinary surgeon.

Salmonella species are non-spore-forming, predominantly motile enterobacteria with cell diameters between about 0.7 and 1.5 μm, lengths from 2 to 5 μm, and peritrichous flagella (all around the cell body). They are chemotrophs, obtaining their energy from oxidation and reduction reactions using organic sources. They are also facultative aerobes, capable of generating ATP with oxygen ("aerobically") when it is available, or when oxygen is not available, using other electron acceptors or fermentation ("anaerobically"). S. enterica subspecies are found worldwide in all warm-blooded animals and in the environment. S. bongori is restricted to cold-blooded animals, particularly reptiles.

Salmonella species are intracellular pathogens; certain serotypes causing illness. Nontyphoidal serotypes can be transferred from animal-to-human and from human-to-human. They usually invade only the gastrointestinal tract and cause salmonellosis, the symptoms of which can be resolved without antibiotics. However, in sub-Saharan Africa, nontyphoidal Salmonella can be invasive and cause paratyphoid fever, which requires immediate treatment with antibiotics. Typhoidal serotypes can only be transferred from human-to-human, and can cause food-borne infection, typhoid fever, and paratyphoid fever. Typhoid fever is caused by Salmonella invading the bloodstream (the typhoidal form), or in addition spreads throughout the body, invades organs, and secretes endotoxins (the septic form). This can lead to life-threatening hypovolemic shock and septic shock, and requires intensive care including antibiotics. The collapse of the Aztec society in Mesoamerica is linked to a catastrophic Salmonella outbreak, one of humanity's deadliest, that occurred after the Spanish conquest.

Taxonomy

The genus Salmonella is part of the family of Enterobacteriaceae. Its taxonomy has been revised and has the potential to confuse. The genus comprises two species, S. bongori and S. enterica, the latter of which is divided into six subspecies: S. e. enterica, S. e. salamae, S. e. arizonae, S. e. diarizonae, S. e. houtenae, and S. e. indica. The taxonomic group contains more than 2500 serotypes (also serovars) defined on the basis of the somatic O (lipopolysaccharide) and flagellar H antigens (the Kauffman–White classification). The full name of a serotype is given as, for example, Salmonella enterica subsp. enterica serotype Typhimurium, but can be abbreviated to Salmonella Typhimurium. Further differentiation of strains to assist clinical and epidemiological investigation may be achieved by antibiotic sensitivity testing and by other molecular biology techniques such as pulsed-field gel electrophoresis, multilocus sequence typing, and, increasingly, whole genome sequencing. Historically, salmonellae have been clinically categorized as invasive (typhoidal) or noninvasive (nontyphoidal salmonellae) based on host preference and disease manifestations in humans.

Serotyping

Serotyping is done by mixing cells with antibodies for a particular antigen. It can give some idea about risk. For example, in 2014, a study showed that S. Reading is very common among young turkey samples, but it is not a significant contributor to human salmonellosis. Serotyping can assist identify the source of contamination. For example, if we have a case of food illness. There are two ways to figure out the source of contamination. One of them is serotyping the Salmonella in infected people, and serotyping Salmonella found at possible source of contamination (e.g. where food was prepared), and then determining if the serotypes match. The other way is to serotype the Salmonella in infected people, see the result and if the result matches with a particular kind of serotype of Salmonella that can only found in a particular species of animals or at particular place, then we can know our source of contamination. Moreover, serotyping may suggest adequate treatment by giving us the idea about the antibiotic resistance of salmonella species. For example, sulfonamides are antibiotics which are appropriate for treatment of Salmonella enterica serotype Typhi (normally it is just shortened to S.Typhi). Serotyping is a faster method than antibiotic sensitivity testing culture methods.

Detection, Culture and Growth Conditions

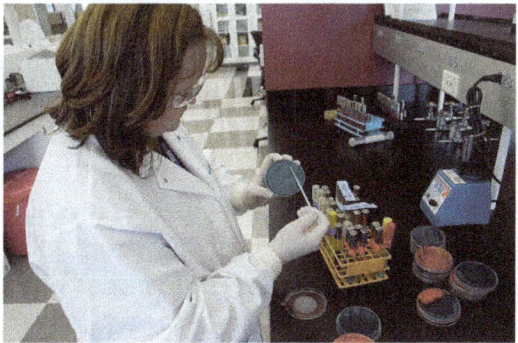

US Food and Drug Administration scientist tests for presence of Salmonella.

Most subspecies of Salmonella produce hydrogen sulfide, which can readily be detected by growing them on media containing ferrous sulfate, such as is used in the triple sugar iron test. Most isolates exist in two phases, a motile phase I and a nonmotile phase II. Cultures that are nonmotile upon primary culture may be switched to the motile phase using a Craigie tube or ditch plate. RVS broth can be used to enrich for Salmonella species for detection in a clinical sample.

Salmonella can also be detected and subtyped using multiplex or real-time polymerase chain reactions (PCR) from extracted Salmonella DNA.

Mathematical models of Salmonella growth kinetics have been developed for chicken, pork, tomatoes, and melons. Salmonella reproduce asexually with a cell division interval of 40 minutes.

Salmonella species lead predominantly host-associated lifestyles, but the bacteria were found to be able to persist in a bathroom setting for weeks following contamination, and are frequently isolated from water sources, which act as bacterial reservoirs and may help to facilitate transmission between hosts. Salmonella is notorious for its ability to survive desiccation and can persist for years in dry environments and foods.

The bacteria are not destroyed by freezing, but UV light and heat accelerate their destruction. They perish after being heated to 55 °C (131 °F) for 90 min, or to 60 °C (140 °F) for 12 min. To protect against Salmonella infection, heating food to an internal temperature of 75 °C (167 °F) is recommended.

Salmonella species can be found in the digestive tracts of humans and animals, especially reptiles. Salmonella on the skin of reptiles or amphibians can be passed to people who handle the animals. Food and water can also be contaminated with the bacteria if they come in contact with the feces of infected people or animals.

Nomenclature

Initially, each Salmonella "species" was named according to clinical considerations, for example Salmonella typhi-murium (mouse typhoid fever), S. cholerae-suis. After host specificity was recognized to not exist for many species, new strains received species names according to the location at which the new strain was isolated. Later, molecular findings led to the hypothesis that Salmonella consisted of only one species, S. enterica, and the serotypes were classified into six groups, two of which are medically relevant. As this now-formalized nomenclature is not in harmony with the traditional usage familiar to specialists in microbiology and infectologists, the traditional nomenclature is still common. Currently, the two recognized species are S. enterica, and S. bongori. In 2005, a third species, Salmonella subterranean, was proposed, but according to the World Health Organization, the bacterium reported does not belong in the genus Salmonella. The six main recognised subspecies are: enterica (serotype I), salamae (serotype II), arizonae (IIIa), diarizonae (IIIb), houtenae (IV), and indica (VI). The former serotype (V) was bongori, which is now considered its own species.

The serotype or serovar, is a classification of Salmonella into subspecies based on antigens that the organism presents. It is based on the Kauffman-White classification scheme that differentiates serological varieties from each other. Serotypes are usually put into subspecies groups after the genus and species, with the serotypes/serovars capitalized, but not italicized: An example is Salmonella enterica serovar Typhimurium.

More modern approaches for typing and subtyping Salmonella include DNA-based methods such as pulsed field gel electrophoresis, multiple-loci VNTR analysis, multilocus sequence typing, and multiplex-PCR-based methods.

Pathogenicity

Salmonella species are facultative intracellular pathogens. A facultative organism uses oxygen to make ATP; when it is not available, it "exercises its option"—the literal meaning of the term—and makes ATP by fermentation, or by substituting one or more of four less efficient electron acceptors as oxygen at the end of the electron transport chain: sulfate, nitrate, sulfur, or fumarate.

Most infections are due to ingestion of food contaminated by animal feces, or by human feces, such as by a food-service worker at a commercial eatery. Salmonella serotypes can be divided into two main groups—typhoidal and nontyphoidal. Nontyphoidal serotypes are more common, and usually cause self-limiting gastrointestinal disease. They can infect a range of animals, and are zoonotic, meaning they can be transferred between humans and other animals. Typhoidal serotypes include Salmonella Typhi and Salmonella Paratyphi A, which are adapted to humans and do not occur in other animals.

Nontyphoidal Salmonella

Non-invasive

Infection with nontyphoidal serotypes of Salmonella generally results in food poisoning. Infection usually occurs when a person ingests foods that contain a high concentration of the bacteria. Infants and young children are much more susceptible to infection, easily achieved by ingesting a small number of bacteria. In infants, infection through inhalation of bacteria-laden dust is possible.

The organisms enter through the digestive tract and must be ingested in large numbers to cause disease in healthy adults. An infection can only begin after living salmonellae (not merely Salmonella-produced toxins) reach the gastrointestinal tract. Some of the microorganisms are killed in the stomach, while the surviving ones enter the small intestine and multiply in tissues. Gastric acidity is responsible for the destruction of the majority of ingested bacteria, but Salmonella has evolved a degree of tolerance to acidic environments that allows a subset of ingested bacteria to survive. Bacterial colonies may also become trapped in mucus produced in the esophagus. By the end of the incubation period, the nearby host cells are poisoned by endotoxins released from the dead salmonellae. The local response to the endotoxins is enteritis and gastrointestinal disorder.

About 2,000 serotypes of nontyphoidal Salmonella are known, which may be responsible for as many as 1.4 million illnesses in the United States each year. People who are at risk for severe illness include infants, elderly, organ-transplant recipients, and the immunocompromised.

Invasive

While in developed countries, nontyphoidal serotypes present mostly as gastrointestinal disease, in sub-Saharan Africa, these serotypes can create a major problem in blood-stream infections, and are the most commonly isolated bacteria from the blood of those presenting with fever. Bloodstream infections caused by nontyphoidal salmonellae in Africa were reported in 2012 to have a case fatality rate of 20–25%. Most cases of invasive nontyphoidal salmonella infection (iNTS) are caused by S. typhimurium or S. enteritidis. A new form of Salmonella typhimurium (ST313) emerged in the southeast of the African continent 75 years ago, followed by a second wave which came out of central Africa 18 years later. This second wave of iNTS possibly originated in the Congo Basin, and early in the event picked up a gene that made it resistant to the antibiotic chloramphenicol. This created the need to use expensive antimicrobial drugs in areas of Africa that were very poor, making treatment difficult. The increased prevalence of iNTS in sub-Saharan Africa compared to other regions is thought to be due to the large proportion of the African population with some degree of immune suppression or impairment due to the burden of HIV, malaria, and malnutrition, especially in children. The genetic makeup of iNTS is evolving into a more typhoid-like bacterium, able to efficiently spread around the human body. Symptoms are reported to be diverse, including fever, hepatosplenomegaly, and respiratory symptoms, often with an absence of gastrointestinal symptoms.

Typhoidal Salmonella

Typhoid fever is caused by Salmonella serotypes which are strictly adapted to humans or higher primates—these include Salmonella typhi, Paratyphi A, Paratyphi B, and Paratyphi C. In the systemic form of the disease, salmonellae pass through the lymphatic system of the intestine into the blood of the patients (typhoid form) and are carried to various organs (liver, spleen, kidneys) to form secondary foci (septic form). Endotoxins first act on the vascular and nervous apparatus, resulting in increased permeability and decreased tone of the vessels, upset of thermal regulation, and vomiting and diarrhoea. In severe forms of the disease, enough liquid and electrolytes are lost to upset the water-salt metabolism, decrease the circulating blood volume and arterial pressure, and cause hypovolemic shock. Septic shock may also develop. Shock of mixed character (with signs of both hypovolemic and septic shock) is more common in severe salmonellosis. Oliguria and azotemia may develop in severe cases as a result of renal involvement due to hypoxia and toxemia.

Molecular Mechanisms of Infection

Mechanisms of infection differ between typhoidal and nontyphoidal serotypes, owing to their different targets in the body and the different symptoms that they cause.

Both groups must enter by crossing the barrier created by the intestinal cell wall, but once they have passed this barrier, they use different strategies to cause infection.

Nontyphoidal serotypes preferentially enter M cells on the intestinal wall by bacterial-mediated endocytosis, a process associated with intestinal inflammation and diarrhoea. They are also able to disrupt tight junctions between the cells of the intestinal wall, impairing the cells' ability to stop the flow of ions, water, and immune cells into and out of the intestine. The combination of the inflammation caused by bacterial-mediated endocytosis and the disruption of tight junctions is thought to contribute significantly to the induction of diarrhoea.

Salmonellae are also able to breach the intestinal barrier via phagocytosis and trafficking by CD18-positive immune cells, which may be a mechanism key to typhoidal Salmonella infection. This is thought to be a more stealthy way of passing the intestinal barrier, and may, therefore, contribute to the fact that lower numbers of typhoidal Salmonella are required for infection than nontyphoidal Salmonella. Salmonella cells are able to enter macrophages via macropinocytosis. Typhoidal serotypes can use this to achieve dissemination throughout the body via the mononuclear phagocyte system, a network of connective tissue that contains immune cells, and surrounds tissue associated with the immune system throughout the body.

Much of the success of Salmonella in causing infection is attributed to two type III secretion systems which function at different times during an infection. One is required for the invasion of nonphagocytic cells, colonization of the intestine, and induction of intestinal inflammatory responses and diarrhea. The other is important for survival in macrophages and establishment of systemic disease. These systems contain many genes which must work co-operatively to achieve infection.

The AvrA toxin injected by the SPI1 type III secretion system of S. Typhimurium works to inhibit the innate immune system by virtue of its serine/threonine acetyltransferase activity, and requires binding to eukaryotic target cell phytic acid (IP6). This leaves the host more susceptible to infection.

Salmonellosis is known to be able to cause back pain or spondylosis. It can manifest as five clinical patterns: gastrointestinal tract infection, enteric fever, bacteremia, local infection, and the chronic reservoir state. The initial symptoms are nonspecific fever, weakness, and myalgia among others. In the bacteremia state, it can spread to any parts of the body and this induces localized infection or it forms abscesses. The forms of localized Salmonella infections are arthritis, urinary tract infection, infection of the central nervous system, bone infection, soft tissue infection, etc. Infection may remain as the latent form for a long time, and when the function of reticular endothelial cells is deteriorated, it may become activated and consequently, it may secondarily induce spreading infection in the bone several months or several years after acute salmonellosis.

Selective Immune Knockout

A 2018 Imperial College London study shows how salmonella disrupt specific arms of the immune system (e.g. 3 of 5 NF-kappaB proteins) using a family of zinc metalloproteinase effectors, leaving others untouched.

Resistance to Oxidative Burst

A hallmark of Salmonella pathogenesis is the ability of the bacterium to survive and proliferate within phagocytes. Phagocytes produce DNA-damaging agents such as nitric oxide and oxygen radicals as a defense against pathogens. Thus, Salmonella species must face attack by molecules that challenge genome integrity. Buchmeier et al. showed that mutants of S. enterica lacking RecA or RecBC protein function are highly sensitive to oxidative compounds synthesized by macrophages, and furthermore these findings indicate that successful systemic infection by S. enterica requires RecA- and RecBC-mediated recombinational repair of DNA damage.

Host Adaptation

S. enterica, through some of its serotypes such as Typhimurium and Enteriditis, shows signs of the ability to infect several different mammalian host species, while other serotypes such as Typhi seem to be restricted to only a few hosts. Some of the ways that Salmonella serotypes have adapted to their hosts include loss of genetic material and mutation. In more complex mammalian species, immune systems, which include pathogen specific immune responses, target serovars of Salmonella through binding of antibodies to structures such as flagella. Through the loss of the genetic material that codes for a flagellum to form, Salmonella can evade a host's immune system. mgtC leader RNA from bacteria virulence gene (mgtCBR operon) decreases flagellin production during infection by directly base pairing with mRNAs of the fljB gene encoding flagellin and promotes degradation. This means that, as these strains of Salmonella have been exposed to similar conditions such as immune systems, similar structures evolved separately to negate these similar, more advanced defenses in hosts. Still, many questions remain about the way that Salmonella has evolved into so many different types, but Salmonella may have evolved through several phases. As Baumler et al. have suggested, Salmonella most likely evolved through horizontal gene transfer, formation of new serovars due to additional pathogenicity islands. and an approximation of its ancestry. So, Salmonella could have evolved into its many different serotypes through gaining genetic information from different pathogenic bacteria. The presence of several pathogenicity islands in the genome of different serotypes has lent credence to this theory.

Salmonella sv. Newport has signs of adaptation to a plant colonization lifestyle, which may play a role in its disproportionate association with foodborne illness linked to produce. A variety of functions selected for during sv. Newport persistence in tomatoes have been reported to be similar to those selected for in sv. Typhimurium from animal hosts. The papA gene, which is unique to sv. Newport, contributes to the strain's fitness in tomatoes, and has homologs in genomes of other Enterobacteriaceae that are able to colonize plant and animal hosts.

Genetics

In addition to its importance as a pathogen, S. enterica serovar Typhimurium has been instrumental in the development of genetic tools that led to an understanding of fundamental bacterial physiology. These developments were enabled by the discovery of the first generalized transducing phage, P22, in Typhimurium that allowed quick and easy genetic exchange that allowed fine structure genetic analysis. The large number of mutants led to a revision of genetic nomenclature for bacteria. Many of the uses of transposons as genetic tools, including transposon delivery, mutagenesis, construction of chromosome rearrangements, were also developed in Typhimurium. These genetic tools also led to a simple test for carcinogens, the Ames test.

PROTEOBACTERIA

Also referred to as "Purple bacteria and relatives", Proteobacteria makes up one of the largest phyla and most versatile phyla in the Bacteria domain. As such, it consists of several types of bacterial that include phototrophs, chemolithotrophs and heterotrophs. Today, well over 460 genera and 1600 species of the phylum have been identified.

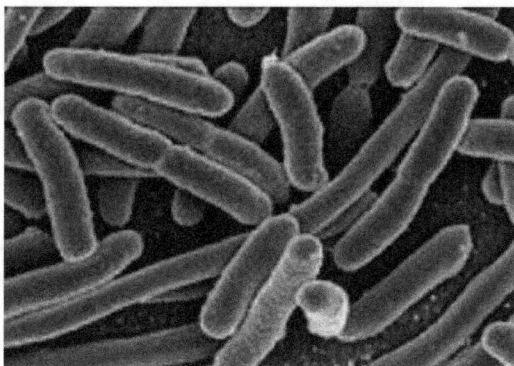

Proteobacteria.

While some are free living, others are pathogenic (plants, animal and human pathogens) and cause a range of diseases.

Examples include:

- Escherichia,

- Vibrio bacteria,

- Salmonella species,

- Helicobacter bacteria.

Phylogenic Groups

Currently, this diverse species are grouped into 6 major phylogenetic groups/classes.

These include:

Alphaproteobacteria

Alphaproteobacteria is the first class and is composed of a variety of bacteria. This includes bacteria of different shapes (rod, spiral, coccobacilli and curved rods etc) as well as differences in metabolism etc.

Despite their differences, Alphaproteobacteria are unified by the fact that they are all oligotrophs. As such, they can survive in environments with low nutrients including glacial ice. This allows them to survive and thrive.

A majority of Alphaproteobacteria are phototrophic which means that they obtain thier energy from sunlight (use sunlight and carbon dioxide to make their own food) due to the presence of photosynthetic pigments.

For the most part, the phototrophic Alphaproteobacteria (purple) have been shown to grow under anoxic conditions where they make their own food in the presence of very little to no oxygen.

Apart from the phototrophic genera in Alphaproteobacteria, this class is also composed of CI-compounds metabolizing bacteria like Methylobacteria species, plant symbionts like Rhizobium species, Nitrifying bacteria like Nitrobacter and animal and human pathogens (Brucella and Arlichia etc). This class of proteobacteria is therefore composed of a wide variety of species.

Whereas some are beneficial to human beings and the environments in general, others are pathogenic and thus cause diseases in human beings and animals. However, some are free living and do not need a host for thier survival.

Phylogenetically, the Alphaproteobacteria are categorized as:

- Aphaproteobacteria (Rhodospirillum and relatives - mostly spiral in shape).

- Alphaproteobacteria (Rhodopseudomonas and relatives - often reproduce through budding).

- Alphaproteobacteria (Rhodobacter and relatives - have carotenoids of spheroidene series and have shown versatility and flexibility in thier metabolism).

Betaproteobacteria

Compared to Alphaproteobacteria, Betaproteobacteria have been shown to require larger amounts of nutrition given that they are eutrophs.

This Gram-negative class is also composed of aerobic and facultative bacteria (some grow in oxygen rich environments while others do not require oxygen for thier survival). Whereas the class Betaproteusbacteria is composed of many chemolithotrophic genera such as Nitrosomonas that oxidize ammonia, it is also composed of various phototrophs and pathogens.

Good examples of pathogenic species include the Neisseriaceae bacteria and the genus Burkholderia. Whereas Neisseriaceae bacteria are responsible for such diseases as gonorrhea and meningitis, members of genus Burkholderia have been associated with such diseases as glanders in horses and other pulmonary infections.

Some of the other genus of Betproteusbacteria includes:

- Bordetella,
- Burkholderia,
- Leptothrix,
- Thiobacillus.

Burkholderia pseudomallei colonies on a Blood agar plate from the CDC Public Health.

Gammaproteobacteria

The class Gammaproteobacteria is one of the largest groups in the bacteria domain. As such, it has been shown to have a large branching phylogenetic tree, which is also one of its distinguishing characteristic. To date, no biochemical trait has been shown to be unique to this class or any of its subgroups.

Despite the diversity that has also been identified in the metabolic activities of Gammaproteobacteria, a majority of the bacteria are chemo-organotrophs that obtain their energy by oxidizing the chemical bonds of various organic compounds.

Apart from chemo-organotrophs, the class is also composed of phototrophs (obtain energy from sunlight) and chemolithotrophs that obtain their metabolic energy from the oxidation of hydrogen and sulfur among others.

Enteric bacteria also make up this class and include organisms like Escherichia coli. These organisms cause diseases in animal and human beings.

Some of the other organisms that belong to the class Gammaproteobacteria include:

- Yersinia pestis,

- Vibrio cholerae,

- Pseudomonas aeruginosa,

- Salmonella.

Deltaproteobacteria is a smaller branch that is largely composed of aerobic genera. Most of these organisms are fruiting body forming that tends to release myxospores when conditions become increasingly unfavorable. The strictly anaerobic genera in this class contain sulfate and sulfur- reducing bacteria.

Anaerobic, sulfate-reducing bacteria include:

- Desulfonema,

- Desulfovibrio,

- Desulfobacter,

- Desulfococcus.

Anaerobic sulfur-reducing bacteria:

- Desulfuromonas chloroethenica,

- Desulfuromonas palmitatis,

- Desulfuromonas thiophila,

- Desulfuromonas svalbardensis.

Some of the characteristics of Myxobacteria include:

- They move by gliding.

- They move in swarms that are known as wolf parks.

- In their swarms, Myxobacteria produce extracellular enzymes that they use to digest food.

- They are social bacteria that can also interact with other bacteria that are outside their own group.

- The fruiting bodies produced by these bacteria are macroscopic and multicellular.

- Myxospores produced by these bacteria are metabolically inactive.

Deltaproteobacteria also consists of a few pathogenic bacteria like SRB Desulfovibrio orale and Bdellovibrio bacteria that live in other gram-negative bacteria as parasites.

In the host bacteria, Bdellovibrio bacteria position themselves in the periplasm where they actively feed on proteins and pollysaccharides. This ultimately kills the host bacteria.

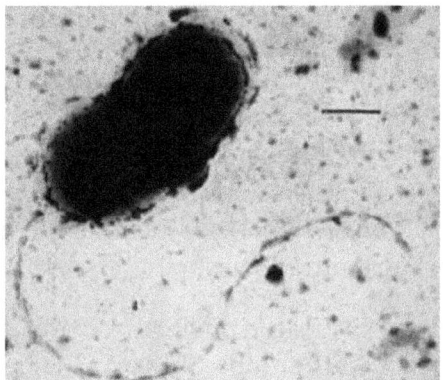

Deltaproteobacteria: Desulfovibrio vulgarism.

Epsilonproteobacteria

Epsilonproteobacteria makes up the smallest class proteobacteria. As such, it is made up of a few genera that are known as microaerophillic bacteria. These are a type of bacteria that require very little levels of oxygen to survive.

Some of the most common Epsilonproteobacteria include species of spirilloid Wolinella, Campylobacter species as well as Helicobacter species.

Campylobacter bacteria and Helicobacter have been associated with a number of diseases that affect human beings and thus make up clinically relevant genera. For instance, whereas Campylobacter bacteria like C. jejuni have been associated with food poisoning that causes severe enteritis in human beings, flagellated bacterium H. Pylori (a member of Helicobacter) have been associated with stomach microbiota and chronic gastritis.

For chemolithotrophic epsilonproteobacteria such as those found in hydrothermal vents and cold seep habitats, energy needs are met by reducing/oxidation of chemical compounds.

Zetaproteobacteria

Zetaproteobacteria is the sixth and most recent class of proteobacteria. A majority of bacteria in this class are found in hydrothermal systems (submarine volcano etc) where they have been shown to contribute to biogeochemical cycling of iron.

Most of the bacteria found in this class are chemolithoautotroph given that they obtain their energy by oxidizing chemical compounds. Ghiorsea bivora is one of the most popular members of Zetaproteobacteria.

Habitats

As mentioned, there are different types of proteobacteria that are classified under six major classes. Within these classes, there are also different types of bacteria that are distinguished by their metabolism mechanisms, shape and habitat etc.

With regards to habitat, this species can be found in a wide range of environments depending on their energy sources and metabolism mechanism. For instance, by looking at members of Alphaproteobacteria, it is possible to identify a wide range of organisms that can be found in different habitats ranging from plant roots (Rhizobium and Azospirillum) inside the cells of a host (rickettsias) and as free living bacteria in the environment such as Methylobacteriaum species.

Being one of the largest and most versatile phyla, therefore, proteobacteria can be found in virtually any environment across the globe. This is made possible by the fact that some of the species in the phyla can survive extreme environments with very little to no oxygen.

Characteristics

Because of their diversity, Proteobacteria have been shown to make up well over 40 percent of all prokaryotic genera of the Gram-negative bacteria.

As prokaryotes, they lack a membrane bound nucleus present in eukaryotes. In addition, they also lack a membrane bound mitochondria and any other membrane bound organelle, which means that most of their organelles are not covered by a membrane.

Because of their diversity, they are also characterized by high morphological and physiological diversity and are composed of rod, cocci and ring shaped bacteria etc.

Morphological characteristics of these organisms play an important role given that it can influence survival in thier environments. In aquatic environments, filamentous proteobacteria can move easily from one area to another and therefore to favorable environments.

While proteobacteria species can be found in an array of environments, a majority of these species are mesophilic, which means that they prefer moderate temperatures (20-45 degrees Celsius) to grow and thrive.

Some have been shown to be thermophilic (such as tepidomonas and thermosulfata that can survive high temperatures) and psychrophilic/cryophiles that are capable of surviving in extremely low temperatures (e.g. Polatomonas).

For a majority of proteobacteria species, motility is made possible by the presence of polar/peritrichous flagella. As a result, a majority of these organisms are motile and can therefore move from one location to another within thier environments.

For such myxobacteria that belong to the Deltaproteobacteria, motility is through gliding. For such bacteria, research studies has also identified highly complex developmental lifestyles as is evident with multicellular structures known as fruiting bodies.

Human Diseases

Some members of Proteobacteria need a host to survive; while some will form a symbiotic relationship with plants and benefit each other, whereas others exist as parasites and thus carry diseases or harm the host. Apart from the different diseases to plants and animals, the phyla is composed of various well known human pathogens that are responsible for intestinal and extra-intestinal diseases.

In human gut, interaction between microbiota and the cells plays an important role in digestion. However, it also helps shape and modulate the immune system. In cases of increased disease (gut) Proteobacteria are often found to have increased in number, which is evidence of instability of microbiota.

Bacteria species belonging to the class Betaproteobacteria. Gammaproteobacteria (V. cholerae and Enterobacteriaceae etc) and Epsilonproteobacteria (C. jejuni etc) have been associated with a variety of intestinal diseases and inflammation and are therefore of great clinical significance.

In addition, these organisms have been associated with such conditions as inflammatory bowel disease (IBD), Crohn's disease (CD) as well as ulcerative colitis (UC) which are chronic infections that are often associated with persistent inflammation of the intestine.

Gram Stain

All proteobacteria are Gram-negative. As such, they do not retain the crystal violet stain (primary stain) during gram staining. This is due to the fact that, like many other gram-negative bacteria, they have a thin peptidoglycan layer that allows gram-positive bacteria to retain the stain.

Some characteristics of the Gram-negative E. coli (member of class Gammaproteobacteria) include:

- Cell wall that is between 5 and 10 micrometers in thickness.

- Monolayer peptidoglycan.

- They lack Teichoic acid on thier cell wall.

- They have a cell envelope outside of thier cell wall.

The following procedure is used in gram staining of Proteobacteria (E. coli):

- Requirements:

  ○ Clean microscope slides,

  ○ Burnsen burner,

  ○ Immersion oil,

  ○ An inoculating loop,

  ○ Distilled water,

  ○ Compound microscope,

  ○ Bacterial sample.

- Reagents:

  ○ Crystal violet,

  ○ Grams Iodine,

  ○ Ethyl alcohol,

  ○ Safranin.

Procedure

- Using a sterilized wire loop, obtain a small sample of the bacteria and make a circular smear in the middle of a clean glass slide.

- Allow the smear to air dray and then heat fix by passing over the Burnsen burner flame - Make sure not to overheat.

- Place the slide on a staining rack and gently flood with the primary stain (crystal violet) for about one minute.

- Tilt the slide and gently wash with distilled water.

- Gently flood the slide with Iodine for about minute and then tilt and gently wash with distilled water.

- Decolorize using ethyl alcohol/acetone for about 5 seconds.

- Rinse with distilled water and gently flood with safranin for about a minute.

- Tilt the slide and rinse with tap water.

- Blot dry to remove any excess water/stain and mount on microscope to observe the cells.

When viewed under the microscope, E. coli cells will appear pink in color. This is because the cells release the primary stain following decolonization. The pink color is a result of the secondary stain (safranin) that is retained (because it is not decolonized).

Benefits

While some members of Proteobacteria are pathogenic and thus carry disease to the host, some are free living and do not need a host for their survival.

There are a variety that present many benefits in their environment. A good example of these bacteria are Nitrogen-fixing bacteria (e.g. Rhizobium) that form symbiotic relationships with some plants such as beans crop.

Here, the bacteria convert atmospheric nitrogen into a form that the plants can use (ammonia). This presents a benefit for the crops and contributes to their growth.

VIBRIO CHOLERAE

Vibrio cholerae is a Gram-negative, comma-shaped bacterium. The bacterium's natural habitat is brackish or saltwater. Some strains of V. cholerae cause the disease cholera. V. cholerae is a facultative anaerobe and has a flagellum at one cell pole as well as pili. V. cholerae can undergo respiratory and fermentative metabolism. When ingested, V. cholerae can cause diarrhoea and vomiting in a host within several hours to 2–3 days of ingestion. V. cholerae was first isolated as the cause of cholera by Italian anatomist Filippo Pacini in 1854, but his discovery was not widely known until Robert Koch, working independently 30 years later, publicized the knowledge and the means of fighting the disease.

Characteristics

V. cholerae is Gram-negative and comma-shaped. Initial isolates are slightly curved, whereas they can appear as straight rods upon laboratory culturing. The bacterium has a flagellum at one cell pole as well as pili. V. cholerae is a facultative anaerobe, and can undergo respiratory and fermentative metabolism. It measures 0.3 micron in diametre and 1.3 micron in length with average swimming velocity of around 75.4 +/- 9.4 microns/sec.

Pathogenesis

V. cholerae pathogenicity genes code for proteins directly or indirectly involved in the virulence of the bacteria. During infection, V. cholerae secretes cholera toxin, a protein that causes profuse, watery diarrhea (known as "rice-water stool"). Colonization of the small intestine also requires the toxin coregulated pilus (TCP), a thin, flexible, filamentous appendage on the surface of bacterial cells. The V. Cholerae particle in the intestinal lumen then uses fimbraie to attach to the intestinal mucosa, not invading the mucosa. After doing so it secretes cholerae toxin causing its symptoms. This then increases cyclic AMP or cAMP by binding (cholerae toxin) to adenylyl cyclase activating the GS pathway which leads to secretion of water into the intestinal lumen causing watery stools or rice watery stools. V. cholerae can cause syndromes ranging from asymptomatic to cholera gravis.

In endemic areas, 75% of cases are asymptomatic, 20% are mild to moderate, and 2-5% are severe forms such as cholera gravis. Symptoms include abrupt onset of watery diarrhea (a grey and cloudy liquid), occasional vomiting, and abdominal cramps. Dehydration ensues, with symptoms and signs such as thirst, dry mucous membranes, decreased skin turgor, sunken eyes, hypotension, weak or absent radial pulse, tachycardia, tachypnea, hoarse voice, oliguria, cramps, renal failure, seizures, somnolence, coma, and death. Death due to dehydration can occur in a few hours to days in untreated children. The disease is also particularly dangerous for pregnant women and their fetuses during late pregnancy, as it may cause premature labor and fetal death. In a study done by the Centers for Disease Control (CDC) in Haiti. They found that women who were pregnant and contracted the disease, 16% of 900 women had fetal death. Risk factors for these deaths include: third trimester, younger maternal age, severe dehydration, and vomiting Dehydration poses the biggest health risk to pregnant women in countries that there are high rates of cholera. In cases of cholera gravis involving severe dehydration, up to 60% of patients can die; however, less than 1% of cases treated with rehydration therapy are fatal. The disease typically lasts 4–6 days. Worldwide, diarrhoeal disease, caused by cholera and many other pathogens, is the second-leading cause of death for children under the age of 5 and at least 120,000 deaths are estimated to be caused by cholera each year. In 2002, the WHO deemed that the case fatality ratio for cholera was about 3.95%.

Disease Occurrence

V. cholerae has an endemic or epidemic occurrence. In countries where the disease has been for the past three years and the cases confirmed are local (within the confines of the country) transmission is considered to be "endemic." Alternatively, an outbreak is declared when the occurrence of disease exceeds the normal occurrence for any given time or location. Epidemics can last several days or over a span of years. Additionally, countries that have an occurrence of an epidemic can also be endemic. The longest standing V. chloerae epidemic was recorded in Yemen. Yemen had two outbreaks, the first occurred between September 2016 and April 2017, and the second began later in

April 2017 and recently was considered to be resolved in 2019 . The epidemic in Yemen took over 2,500 lives and impacted over 1 million people of Yemen.

Preventative Measures

When visiting areas with epidemic cholera, the following precautions should be observed: drink and use bottled water; frequently wash hands with soap and safe water; use chemical toilets or bury feces if no restroom is available; do not defecate in any body of water and cook food thoroughly. Hand hygiene is an essential in areas where soap and water is not available. When there is no sanitation available for hand washing, scrub hands with ash or sand and rinse with clean water. A single dose vaccine is available for those traveling to an area where cholera is common.

There is a V. cholerae vaccine available to prevent disease spread. The vaccine is known as the, "oral cholera vaccine" (OCV). There are three types of OCV available for prevention: Dukoral, Shanchol, and Euvichol-Plus. All three OCVs require two doses to be fully effective. Countries who are endemic or have an epidemic status are eligible to receive the vaccine based on several criteria: Risk of cholera, Severity of cholera, WASH conditions and capacity to improve, Healthcare conditions and capacity to improve, Capacity to implement OCV campaigns, Capacity to conduct M&E activities, Commitment at national and local level Since May the start of the OCV program to May 2018 over 25 million vaccines have been distributed to countries who meet the above criteria.

Genome

V. cholerae has two circular chromosomes, together totalling 4 million base pairs of DNA sequence and 3,885 predicted genes. The genes for cholera toxin are carried by CTX-phi (CTXφ), a temperate bacteriophage inserted into the V. cholerae genome. CTXφ can transmit cholera toxin genes from one V. cholerae strain to another, one form of horizontal gene transfer. The genes for toxin coregulated pilus are coded by the Vibrio pathogenicity island (VPI). The entire genome of the virulent strain V. cholerae El Tor N16961 has been sequenced, and contains two circular chromosomes. Chromosome 1 has 2,961,149 base pairs with 2,770 open reading frames (ORF's) and chromosome 2 has 1,072,315 base pairs, 1,115 ORF's. The larger first chromosome contains the crucial genes for toxicity, regulation of toxicity, and important cellular functions, such as transcription and translation.

The second chromosome is determined to be different from a plasmid or megaplasmid due to the inclusion of housekeeping and other essential genes in the genome, including essential genes for metabolism, heat-shock proteins, and 16S rRNA genes, which are ribosomal subunit genes used to track evolutionary relationships between bacteria. Also relevant in determining if the replicon is a chromosome is whether it represents a significant percentage of the genome, and chromosome 2 is 40% by size of the entire genome. And, unlike plasmids, chromosomes are not self-transmissible. However, the second chromosome may have once been a megaplasmid because it contains some genes usually found on plasmids.

V. cholerae contains a genomic island of pathogenicity and is lysogenized with phage DNA. That means that the genes of a virus were integrated into the bacterial genome and made the bacteria pathogenic. The molecular pathway involved in expression of virulence.

Bacteriophage CTXφ

CTXφ (also called CTXphi) is a filamentous phage that contains the genes for cholera toxin. Infectious CTXφ particles are produced when V. cholerae infects humans. Phage particles are secreted from bacterial cells without lysis. When CTXφ infects V. cholerae cells, it integrates into specific sites on either chromosome. These sites often contain tandem arrays of integrated CTXφ prophage. In addition to the ctxA and ctxB genes encoding cholera toxin, CTXφ contains eight genes involved in phage reproduction, packaging, secretion, integration, and regulation. The CTXφ genome is 6.9 kb long.

Vibrio Pathogenicity Island

The Vibrio pathogenicity island (VPI) contains genes primarily involved in the production of toxin coregulated pilus (TCP). It is a large genetic element (about 40 kb) flanked by two repetitive regions (att-like sites), resembling a phage genome in structure. The VPI contains two gene clusters, the TCP cluster, and the ACF cluster, along with several other genes. The acf cluster is composed of four genes: acfABCD. The tcp cluster is composed of 15 genes: tcpABCDEFHIJPQRST and regulatory gene toxT.

Ecology and Epidemiology

The main reservoirs of V. cholerae are people and aquatic sources such as brackish water and estuaries, often in association with copepods or other zooplankton, shellfish, and aquatic plants.

Cholera infections are most commonly acquired from drinking water in which V. cholerae is found naturally or into which it has been introduced from the feces of an infected person. Other common vehicles include contaminated fish and shellfish, produce, or leftover cooked grains that have not been properly reheated. Transmission from person to person, even to health care workers during epidemics, is rarely documented. V. cholerae thrives in a aquatic environment, particularly in surface water. The primary connection between humans and pathogenic strains is through water, particularly in economically reduced areas that do not have good water purification systems.

Nonpathogenic strains are also present in water ecologies. The wide variety of pathogenic and nonpathogenic strains that co-exist in aquatic environments are thought to allow for so many genetic varieties. Gene transfer is fairly common amongst bacteria, and recombination of different V. cholerae genes can lead to new virulent strains.

A symbiotic relationship between V. cholerae and Ruminococcus obeum has been determined. R. obeum autoinducer represses the expression of several V. cholerae virulence factors. This inhibitory mechanism is likely to be present in other gut microbiota species which opens the way to mine the gut microbiota of members in specific communities which may utilize autoinducers or other mechanisms in order to restrict colonization by V. cholerae or other enteropathogens.

Diversity and Evolution

Two serogroups of V. cholerae, O1 and O139, cause outbreaks of cholera. O1 causes the majority of outbreaks, while O139 – first identified in Bangladesh in 1992 – is confined to Southeast Asia. Many other serogroups of V. cholerae, with or without the cholera toxin gene (including the nontoxigenic strains of the O1 and O139 serogroups), can cause a cholera-like illness. Only toxigenic strains of serogroups O1 and O139 have caused widespread epidemics.

V. cholerae O1 has two biotypes, classical and El Tor, and each biotype has two distinct serotypes, Inaba and Ogawa. The symptoms of infection are indistinguishable, although more people infected with the El Tor biotype remain asymptomatic or have only a mild illness. In recent years, infections with the classical biotype of V. cholerae O1 have become rare and are limited to parts of Bangladesh and India. Recently, new variant strains have been detected in several parts of Asia and Africa. Observations suggest these strains cause more severe cholera with higher case fatality rates.

Natural Genetic Transformation

V. cholerae can be induced to become competent for natural genetic transformation when grown on chitin, a biopolymer that is abundant in aquatic habitats (e.g. from crustacean exoskeletons). Natural genetic transformation is a sexual process involving DNA transfer from one bacterial cell to another through the intervening medium, and the integration of the donor sequence into the recipient genome by homologous recombination. Transformation competence in V. cholerae is stimulated by increasing cell density accompanied by nutrient limitation, a decline in growth rate, or stress. The V. cholerae uptake machinery involves a competence-induced pilus, and a conserved DNA binding protein that acts as a ratchet to reel DNA into the cytoplasm.

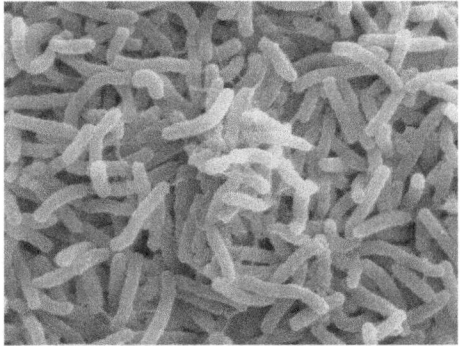

Vibrio cholerae bacteria.

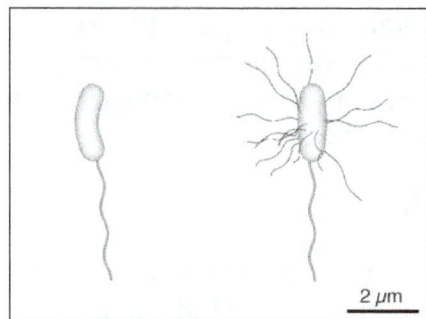

Diagram of the bacterium, V. cholerae.

Microscope slide with a sample of "colera asiaticus", prepared by Pacini.

PATHOGENIC BACTERIA

Pathogenic bacteria are bacteria that can cause disease. Although most bacteria are harmless or often beneficial, some are pathogenic, with the number of species estimated as fewer than a hundred that are seen to cause infectious diseases in humans. By contrast, several thousand species exist in the human digestive system.

One of the bacterial diseases with the highest disease burden is tuberculosis, caused by Mycobacterium tuberculosis bacteria, which kills about 2 million people a year, mostly in sub-Saharan Africa. Pathogenic bacteria contribute to other globally important diseases, such as pneumonia, which can be caused by bacteria such as Streptococcus and Pseudomonas, and foodborne illnesses, which can be caused by bacteria such as Shigella, Campylobacter, and Salmonella. Pathogenic bacteria also cause infections such as tetanus, typhoid fever, diphtheria, syphilis, and leprosy. Pathogenic bacteria are also the cause of high infant mortality rates in developing countries. Koch's postulates are the standard to establish a causative relationship between a microbe and a disease.

Diseases

Each species has specific effect and causes symptoms in people who are infected. Some, if not most people who are infected with a pathogenic bacteria do not have symptoms. Immuno-compromised individuals are more susceptible to pathogenic bacteria.

Pathogenic Susceptibility

Some pathogenic bacteria cause disease under certain conditions, such as entry through the skin via a cut, through sexual activity or through a compromised immune function.

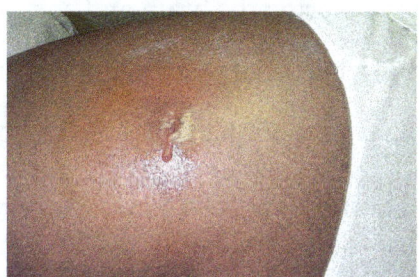

An abscess caused by opportunistic S. aureus bacteria.

Streptococcus and Staphylococcus are part of the normal skin microbiota and typically reside on healthy skin or in the nasopharangeal region. Yet these species can potentially initiate skin infections. They are also able to cause sepsis, pneumonia or meningitis. These infections can become quite serious creating a systemic inflammatory response resulting in massive vasodilation, shock, and death.

Other bacteria are opportunistic pathogens and cause disease mainly in people suffering from immunosuppression or cystic fibrosis. Examples of these opportunistic pathogens include Pseudomonas aeruginosa, Burkholderia cenocepacia, and Mycobacterium avium.

Intracellular

Obligate intracellular parasites (e.g. Chlamydophila, Ehrlichia, Rickettsia) have the ability to only grow and replicate inside other cells. Even these intracellular infections may be asymptomatic, requiring an incubation period. An example of this is Rickettsia which causes typhus. Another causes Rocky Mountain spotted fever.

Chlamydia is a phylum of intracellular parasites. These pathogens can cause pneumonia or urinary tract infection and may be involved in coronary heart disease.

Other groups of intracellular bacterial pathogens include Salmonella, Neisseria, Brucella, Mycobacterium, Nocardia, Listeria, Francisella, Legionella, and Yersinia pestis. These can exist intracellularly, but can exist outside of host cells.

Infections in Specific Tissue

Bacterial pathogens often cause infection in specific areas of the body. Others are generalists.

- Bacterial vaginosis is caused by bacteria that change the vaginal microbiota caused by an overgrowth of bacteria that crowd out the Lactobacilli species that maintain healthy vaginal microbial populations.

- Other non-bacterial vaginal infections include: yeast infection (candidiasis), *Trichomonas vaginalis* (trichomoniasis).

- Bacterial meningitis is a bacterial inflammation of the meninges, that is, the protective membranes covering the brain and spinal cord.

- Bacterial pneumonia is a bacterial infection of the lungs.

- Urinary tract infection is predominantly caused by bacteria. Symptoms include the strong and frequent sensation or urge to urinate, pain during urination, and urine that is cloudy. The main causal agent is *Escherichia coli*. Urine is typically sterile but contains a variety of salts, and waste products. Bacteria can ascend into the bladder or kidney and causing cystitis and nephritis.

- Bacterial gastroenteritis is caused by enteric, pathogenic bacteria. These pathogenic species are usually distinct from the usually harmless bacteria of the normal gut flora. But a different strain of the same species may be pathogenic. The distinction is sometimes difficult as in the case of *Escherichia*.

- Bacterial skin infections include:

  ○ Impetigo is a highly contagious bacterial skin infection commonly seen in children. It is caused by *Staphylococcus aureus*, and *Streptococcus pyogenes*.

  ○ Erysipelas is an acute streptococcus bacterial infection of the deeper skin layers that spreads via with lymphatic system.

  ○ Cellulitis is a diffuse inflammation of connective tissue with severe inflammation of dermal and subcutaneous layers of the skin. Cellulitis can be caused by normal skin flora or by contagious contact, and usually occurs through open skin, cuts, blisters, cracks in the skin, insect bites, animal bites, burns, surgical wounds, intravenous drug injection, or sites of intravenous catheter insertion. In most cases it is the skin on the face or lower legs that is affected, though cellulitis can occur in other tissues.

Mechanisms of Damage

The symptoms of disease appear as pathogenic bacteria damage host tissues or interfere with their function. The bacteria can damage host cells directly. They can also cause damage indirectly by provoking an immune response that inadvertently damages host cells.

Direct

Once pathogens attach to host cells, they can cause direct damage as the pathogens use the host cell for nutrients and produce waste products. For example, Streptococcus mutans, a component of dental plaque, metabolizes dietary sugar and produces acid as a waste product. The acid decalcifies the tooth surface to cause dental caries. However, toxins produced by bacteria cause most of the direct damage to host cells.

Toxin Production

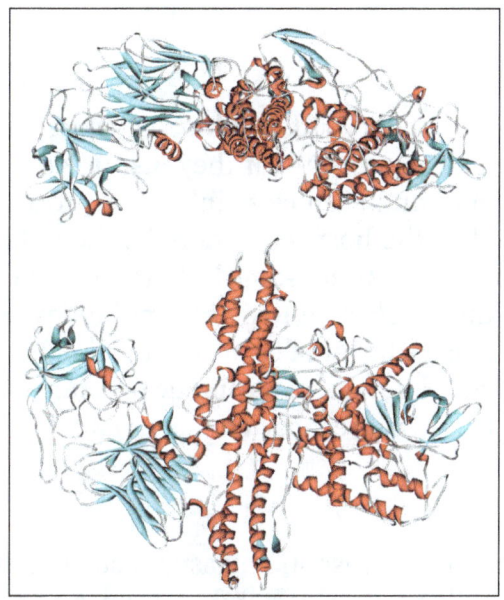

Protein structure of Botulinum toxin 3BTA.

Endotoxins are the lipid portions of lipopolysaccharides that are part of the outer membrane of the cell wall of gram negative bacteria. Endotoxins are released when the bacteria lyses, which is why after antibiotic treatment, symptoms can worsen at first as the bacteria are killed and they release their endotoxins. Exotoxins are secreted into the surrounding medium or released when the bacteria die and the cell wall breaks apart.

Indirect

An excessive or inappropriate immune response triggered by an infection may damage host cells.

Survival in Host

Nutrients

Iron is required for humans, as well as the growth of most bacteria. To obtain free iron, some pathogens secrete proteins called siderophores, which take the iron away from iron-transport proteins by binding to the iron even more tightly. Once the iron-siderophore complex is formed, it is taken up by siderophore receptors on the bacterial surface and then that iron is brought into the bacterium.

Identification

Typically identification is done by growing the organism in a wide range of cultures which can take up to 48 hours. The growth is then visually or genomically identified. The cultured organism is then subjected to various assays to observe reactions to help further identify species and strain.

Treatment

Bacterial infections may be treated with antibiotics, which are classified as bacterio-cidal if they kill bacteria or bacteriostatic if they just prevent bacterial growth. There are many types of antibiotics and each class inhibits a process that is different in the pathogen from that found in the host. For example, the antibiotics chloramphenicol and tetracyclin inhibit the bacterial ribosome but not the structurally different eukary-otic ribosome, so they exhibit selective toxicity. Antibiotics are used both in treating human disease and in intensive farming to promote animal growth. Both uses may be contributing to the rapid development of antibiotic resistance in bacterial populations. Phage therapy can also be used to treat certain bacterial infections.

Prevention

Infections can be prevented by antiseptic measures such as sterilizing the skin prior to piercing it with the needle of a syringe and by proper care of indwelling catheters. Surgical and dental instruments are also sterilized to prevent infection by bacteria. Dis-infectants such as bleach are used to kill bacteria or other pathogens on surfaces to pre-vent contamination and further reduce the risk of infection. Bacteria in food are killed by cooking to temperatures above 73 °C (163 °F).

ESCHERICHIA COLI

Escherichia coli, also known as E. coli is a Gram-negative, facultative anaerobic, rod-shaped, coliform bacterium of the genus Escherichia that is commonly found in the lower intestine of warm-blooded organisms (endotherms). Most E. coli strains are harmless, but some serotypes can cause serious food poisoning in their hosts, and are occasionally responsible for product recalls due to food contamination. The harmless strains are part of the normal microbiota of the gut, and can benefit their hosts by pro-ducing vitamin K_2, and preventing colonization of the intestine with pathogenic bac-teria, having a symbiotic relationship. E. coli is expelled into the environment within fecal matter. The bacterium grows massively in fresh fecal matter under aerobic condi-tions for 3 days, but its numbers decline slowly afterwards.

E. coli and other facultative anaerobes constitute about 0.1% of gut microbiota, and fecal–oral transmission is the major route through which pathogenic strains of the bacterium cause disease. Cells are able to survive outside the body for a limited amount of time, which makes them potential indicator organisms to test environmental samples for fecal contamination. A growing body of research, though, has examined environmentally persistent E. coli which can survive for extended periods outside a host.

The bacterium can be grown and cultured easily and inexpensively in a laboratory setting, and has been intensively investigated for over 60 years. E. coli is a chemoheterotroph whose chemically defined medium must include a source of carbon and energy. E. coli is the most widely studied prokaryotic model organism, and an important species in the fields of biotechnology and microbiology, where it has served as the host organism for the majority of work with recombinant DNA. Under favorable conditions, it takes up to 20 minutes to reproduce.

Biology and Biochemistry

Type and Morphology

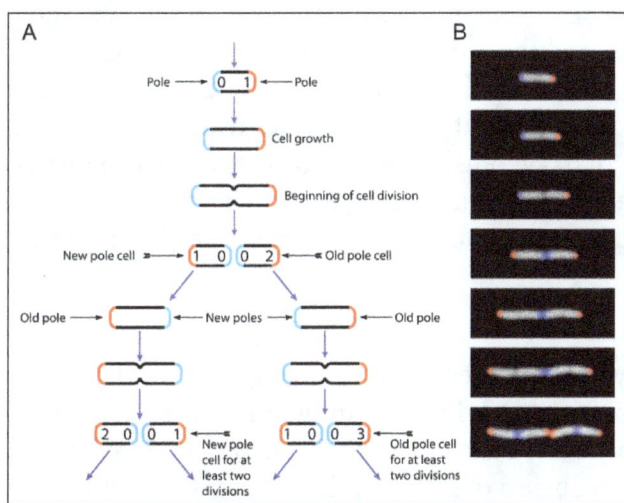

Model of successive binary fission in E. coli.

E. coli is a Gram-negative, facultative anaerobe (that makes ATP by aerobic respiration if oxygen is present, but is capable of switching to fermentation or anaerobic respiration if oxygen is absent) and nonsporulating bacterium. Cells are typically rod-shaped, and are about 2.0 μm long and 0.25–1.0 μm in diameter, with a cell volume of 0.6–0.7 μm³.

E. coli stains Gram-negative because its cell wall is composed of a thin peptidoglycan layer and an outer membrane. During the staining process, E. coli picks up the color of the counterstain safranin and stains pink. The outer membrane surrounding the cell wall provides a barrier to certain antibiotics such that E. coli is not damaged by penicillin.

Strains that possess flagella are motile. The flagella have a peritrichous arrangement. It also attaches and effaces to the microvilli of the intestines via an adhesion molecule known as intimin.

Metabolism

E. coli can live on a wide variety of substrates and uses mixed-acid fermentation in anaerobic conditions, producing lactate, succinate, ethanol, acetate, and carbon dioxide. Since many pathways in mixed-acid fermentation produce hydrogen gas, these pathways require the levels of hydrogen to be low, as is the case when E. coli lives together with hydrogen-consuming organisms, such as methanogens or sulphate-reducing bacteria.

Culture Growth

Optimum growth of E. coli occurs at 37 °C (98.6 °F), but some laboratory strains can multiply at temperatures up to 49 °C (120 °F). E. coli grows in a variety of defined laboratory media, such as lysogeny broth, or any medium that contains glucose, ammonium phosphate monobasic, sodium chloride, magnesium sulfate, potassium phosphate dibasic, and water. Growth can be driven by aerobic or anaerobic respiration, using a large variety of redox pairs, including the oxidation of pyruvic acid, formic acid, hydrogen, and amino acids, and the reduction of substrates such as oxygen, nitrate, fumarate, dimethyl sulfoxide, and trimethylamine N-oxide. E. coli is classified as a facultative anaerobe. It uses oxygen when it is present and available. It can, however, continue to grow in the absence of oxygen using fermentation or anaerobic respiration. The ability to continue growing in the absence of oxygen is an advantage to bacteria because their survival is increased in environments where water predominates.

Cell Cycle

The bacterial cell cycle is divided into three stages. The B period occurs between the completion of cell division and the beginning of DNA replication. The C period encompasses the time it takes to replicate the chromosomal DNA. The D period refers to the stage between the conclusion of DNA replication and the end of cell division. The doubling rate of E. coli is higher when more nutrients are available. However, the length of the C and D periods do not change, even when the doubling time becomes less than the sum of the C and D periods. At the fastest growth rates, replication begins before the previous round of replication has completed, resulting in multiple replication forks along the DNA and overlapping cell cycles.

Genetic Adaptation

E. coli and related bacteria possess the ability to transfer DNA via bacterial conjugation or transduction, which allows genetic material to spread horizontally through an existing population. The process of transduction, which uses the bacterial virus called a bacteriophage, is where the spread of the gene encoding for the Shiga toxin from the Shigella bacteria to E. coli helped produce E. coli O157:H7, the Shiga toxin-producing strain of E. coli.

Diversity

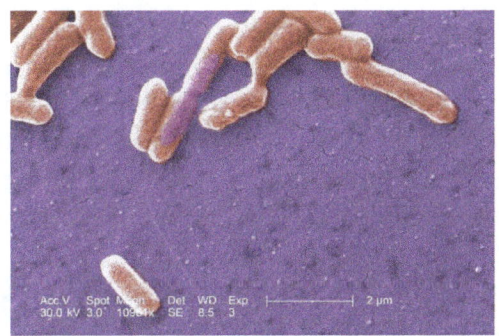

Scanning electron micrograph of an E. coli colony.

E. coli encompasses an enormous population of bacteria that exhibit a very high degree of both genetic and phenotypic diversity. Genome sequencing of a large number of isolates of E. coli and related bacteria shows that a taxonomic reclassification would be desirable. However, this has not been done, largely due to its medical importance, and E. coli remains one of the most diverse bacterial species: only 20% of the genes in a typical E. coli genome is shared among all strains.

In fact, from the evolutionary point of view, the members of genus Shigella (S. dysenteriae, S. flexneri, S. boydii, and S. sonnei) should be classified as E. coli strains, a phenomenon termed taxa in disguise. Similarly, other strains of E. coli (e.g. the K-12 strain commonly used in recombinant DNA work) are sufficiently different that they would merit reclassification.

A strain is a subgroup within the species that has unique characteristics that distinguish it from other strains. These differences are often detectable only at the molecular level; however, they may result in changes to the physiology or lifecycle of the bacterium. For example, a strain may gain pathogenic capacity, the ability to use a unique carbon source, the ability to take upon a particular ecological niche, or the ability to resist antimicrobial agents. Different strains of E. coli are often host-specific, making it possible to determine the source of fecal contamination in environmental samples. For example, knowing which E. coli strains are present in a water sample allows researchers to make assumptions about whether the contamination originated from a human, another mammal, or a bird.

Serotypes

A common subdivision system of E. coli, but not based on evolutionary relatedness, is by serotype, which is based on major surface antigens (O antigen: part of lipopolysaccharide layer; H: flagellin; K antigen: capsule), e.g. O157:H7). It is, however, common to cite only the serogroup, i.e. the O-antigen. At present, about 190 serogroups are known. The common laboratory strain has a mutation that prevents the formation of an O-antigen and is thus not typeable.

Genome Plasticity and Evolution

Like all lifeforms, new strains of E. coli evolve through the natural biological processes of mutation, gene duplication, and horizontal gene transfer; in particular, 18% of the genome of the laboratory strain MG1655 was horizontally acquired since the divergence from Salmonella. E. coli K-12 and E. coli B strains are the most frequently used varieties for laboratory purposes. Some strains develop traits that can be harmful to a host animal. These virulent strains typically cause a bout of diarrhea that is often self-limiting in healthy adults but is frequently lethal to children in the developing world. More virulent strains, such as O157:H7, cause serious illness or death in the elderly, the very young, or the immunocompromised.

The genera Escherichia and Salmonella diverged around 102 million years ago (credibility interval: 57–176 mya), which coincides with the divergence of their hosts: the former being found in mammals and the latter in birds and reptiles. This was followed by a split of an Escherichia ancestor into five species (E. albertii, E. coli, E. fergusonii, E. hermannii, and E. vulneris). The last E. coli ancestor split between 20 and 30 million years ago.

The long-term evolution experiments using E. coli, begun by Richard Lenski in 1988, have allowed direct observation of genome evolution over more than 65,000 generations in the laboratory. For instance, E. coli typically do not have the ability to grow aerobically with citrate as a carbon source, which is used as a diagnostic criterion with which to differentiate E. coli from other, closely, related bacteria such as Salmonella. In this experiment, one population of E. coli unexpectedly evolved the ability to aerobically metabolize citrate, a major evolutionary shift with some hallmarks of microbial speciation.

Neotype Strain

E. coli is the type species of the genus (Escherichia) and in turn Escherichia is the type genus of the family Enterobacteriaceae, where the family name does not stem from the genus Enterobacter + "i" (sic.) + "aceae", but from "enterobacterium" + "aceae" (enterobacterium being not a genus, but an alternative trivial name to enteric bacterium).

The original strain described by Escherich is believed to be lost, consequently a new type strain (neotype) was chosen as a representative: the neotype strain is U5/41T, also known under the deposit names DSM 30083, ATCC 11775, and NCTC 9001, which is pathogenic to chickens and has an O1:K1:H7 serotype. However, in most studies, either O157:H7, K-12 MG1655, or K-12 W3110 were used as a representative E. coli. The genome of the type strain has only lately been sequenced.

Phylogeny of E. Coli Strains

A large number of strains belonging to this species have been isolated and characterised. In addition to serotype (vide supra), they can be classified according to their phylogeny, i.e. the inferred evolutionary history. Particularly the use of whole genome sequences yields highly supported phylogenies. Based on such data, five subspecies of E. coli were distinguished.

The link between phylogenetic distance ("relatedness") and pathology is small, e.g. the O157:H7 serotype strains, which form a clade ("an exclusive group")—group E—are all enterohaemorragic strains (EHEC), but not all EHEC strains are closely related. In fact, four different species of Shigella are nested among E. coli strains (vide supra), while E. albertii and E. fergusonii are outside this group. Indeed, all Shigella species were placed within a single subspecies of E. coli in a phylogenomic study that included the type strain, and for this reason an according reclassification is difficult. All commonly used research strains of E. coli belong to group A and are derived mainly from Clifton's K-12 strain (λ^+ F^+; O16) and to a lesser degree from d'Herelle's Bacillus coli strain (B strain)(O7).

Genomics

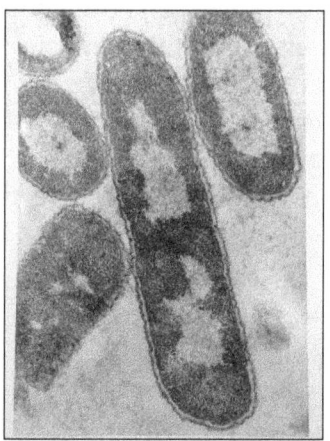

An image of E.coli using early electron microscopy.

The first complete DNA sequence of an *E. coli* genome (laboratory strain K-12 derivative MG1655) was published in 1997. It is a circular DNA molecule 4.6 million base pairs in length, containing 4288 annotated protein-coding genes (organized into 2584 operons), seven ribosomal RNA (rRNA) operons, and 86 transfer RNA (tRNA) genes. Despite having been the subject of intensive genetic analysis for about 40 years, a large number of these genes were previously unknown. The coding density was found to be very high, with a mean distance between genes of only 118 base pairs. The genome was observed to contain a significant number of transposable genetic elements, repeat elements, cryptic prophages, and bacteriophage remnants.

More than three hundred complete genomic sequences of *Escherichia* and *Shigella* species are known. The genome sequence of the type strain of *E. coli* was added to this collection before 2014. Comparison of these sequences shows a remarkable amount of diversity; only about 20% of each genome represents sequences present in every one of the isolates, while around 80% of each genome can vary among isolates. Each individual genome contains between 4,000 and 5,500 genes, but the total number of different genes among all of the sequenced *E. coli* strains (the pangenome) exceeds 16,000. This very large variety of component genes has been interpreted to mean that two-thirds of the *E. coli* pangenome originated in other species and arrived through the process of horizontal gene transfer.

Gene Nomenclature

Genes in *E. coli* are usually named by 4-letter acronyms that derive from their function (when known) and italicized. For instance, *recA* is named after its role in homologous recombination plus the letter A. Functionally related genes are named *recB*, *recC*, *recD* etc. The proteins are named by uppercase acronyms, e.g. RecA, RecB, etc. When the genome of *E. coli* was sequenced, all genes were numbered (more or less) in their order on the genome and abbreviated by b numbers, such as b2819 (= *recD*). The "b" names were created after Fred Blattner, who led the genome sequence effort. Another numbering system was introduced with the sequence of another *E. coli* strain, W3110, which was sequenced in Japan and hence uses numbers starting by JW (Japanese W3110), e.g. JW2787 (= *recD*). Hence, *recD* = b2819 = JW2787. Note, however, that most databases have their own numbering system, e.g. the EcoGene database uses EG10826 for *recD*. Finally, ECK numbers are specifically used for alleles in the MG1655 strain of *E. coli* K-12. Complete lists of genes and their synonyms can be obtained from databases such as EcoGene or Uniprot.

Proteomics

Proteome

Several studies have investigated the proteome of E. coli. By 2006, 1,627 (38%) of the 4,237 open reading frames (ORFs) had been identified experimentally. The 4,639,221–base pair sequence of Escherichia coli K-12 is presented. Of 4288 protein-coding genes annotated, 38 percent have no attributed function. Comparison with five other sequenced microbes reveals ubiquitous as well as narrowly distributed gene families; many families of similar genes within E. coli are also evident. The largest family of paralogous proteins contains 80 ABC transporters. The genome as a whole is strikingly organized with respect to the local direction of replication; guanines, oligonucleotides possibly related to replication and recombination, and most genes are so oriented. The genome also contains insertion sequence (IS) elements, phage remnants, and many other patches of unusual composition indicating genome plasticity through horizontal transfer.

Interactome

The interactome of E. coli has been studied by affinity purification and mass spectrometry (AP/MS) and by analyzing the binary interactions among its proteins.

Protein complexes: A 2006 study purified 4,339 proteins from cultures of strain K-12 and found interacting partners for 2,667 proteins, many of which had unknown functions at the time. A 2009 study found 5,993 interactions between proteins of the same E. coli strain, though these data showed little overlap with those of the 2006 publication.

Binary interactions: Rajagopala et al. have carried out systematic yeast two-hybrid screens with most E. coli proteins, and found a total of 2,234 protein-protein interactions. This study also integrated genetic interactions and protein structures and mapped 458 interactions within 227 protein complexes.

Normal Microbiota

E. coli belongs to a group of bacteria informally known as coliforms that are found in the gastrointestinal tract of warm-blooded animals. E. coli normally colonizes an infant's gastrointestinal tract within 40 hours of birth, arriving with food or water or from the individuals handling the child. In the bowel, E. coli adheres to the mucus of the large intestine. It is the primary facultative anaerobe of the human gastrointestinal tract. (Facultative anaerobes are organisms that can grow in either the presence or absence of oxygen.) As long as these bacteria do not acquire genetic elements encoding for virulence factors, they remain benign commensals.

Therapeutic Use

Nonpathogenic E. coli strain Nissle 1917, (Mutaflor) and E. coli O83:K24:H31 (Colinfant)) are used as probiotic agents in medicine, mainly for the treatment of various gastrointestinal diseases, including inflammatory bowel disease.

Role in Disease

Most E. coli strains do not cause disease, naturally living in the gut, but virulent strains can cause gastroenteritis, urinary tract infections, neonatal meningitis, hemorrhagic colitis, and Crohn's disease. Common signs and symptoms include severe abdominal cramps, diarrhea, hemorrhagic colitis, vomiting, and sometimes fever. In rarer cases, virulent strains are also responsible for bowel necrosis (tissue death) and perforation without progressing to hemolytic-uremic syndrome, peritonitis, mastitis, sepsis, and Gram-negative pneumonia. Very young children are more susceptible to develop severe illness, such as hemolytic uremic syndrome; however, healthy individuals of all ages are at risk to the severe consequences that may arise as a result of being infected with E. coli.

Some strains of E. coli, for example O157:H7, can produce Shiga toxin (classified as a bioterrorism agent). The Shiga toxin causes inflammatory responses in target cells of the gut, leaving behind lesions which result in the bloody diarrhea that is a symptom of a Shiga toxin-producing E. coli (STEC) infection. This toxin further causes premature destruction of the red blood cells, which then clog the body's filtering system, the kidneys, in some rare cases (usually in children and the elderly) causing hemolytic-uremic syndrome (HUS), which may lead to kidney failure and even death. Signs of hemolytic uremic syndrome include decreased frequency of urination, lethargy, and paleness of cheeks and inside the lower eyelids. In 25% of HUS patients, complications of nervous system occur, which in turn causes strokes. In addition, this strain causes the buildup of fluid, leading to edema around the lungs and legs and arms. This increase in fluid buildup especially around the lungs impedes the functioning of the heart, causing an increase in blood pressure.

Uropathogenic E. coli (UPEC) is one of the main causes of urinary tract infections. It is part of the normal microbiota in the gut and can be introduced in many ways. In particular for females, the direction of wiping after defecation (wiping back to front) can lead to fecal contamination of the urogenital orifices. Anal intercourse can also introduce this bacterium into the male urethra, and in switching from anal to vaginal intercourse, the male can also introduce UPEC to the female urogenital system.

Enterotoxigenic E. coli (ETEC) is the most common cause of traveler's diarrhea, with as many as 840 million cases worldwide in developing countries each year. The bacteria, typically transmitted through contaminated food or drinking water, adheres to the intestinal lining, where it secretes either of two types of enterotoxins, leading to watery diarrhea. The rate and severity of infections are higher among children under the age of five, including as many as 380,000 deaths annually.

In May 2011, one E. coli strain, O104:H4, was the subject of a bacterial outbreak that began in Germany. Certain strains of E. coli are a major cause of foodborne illness. The outbreak started when several people in Germany were infected with enterohemorrhagic E. coli (EHEC) bacteria, leading to hemolytic-uremic syndrome (HUS), a medical emergency that requires urgent treatment. The outbreak did not only concern Germany, but also 15 other countries, including regions in North America. On 30 June 2011, the German Bundesinstitut für Risikobewertung (BfR) (Federal Institute for Risk Assessment, a federal institute within the German Federal Ministry of Food, Agriculture and Consumer Protection) announced that seeds of fenugreek from Egypt were likely the cause of the EHEC outbreak.

Incubation Period

The time between ingesting the STEC bacteria and feeling sick is called the "incubation period". The incubation period is usually 3–4 days after the exposure, but may be as short as 1 day or as long as 10 days. The symptoms often begin slowly with mild belly pain or non-bloody diarrhea that worsens over several days. HUS, if it occurs, develops an average 7 days after the first symptoms, when the diarrhea is improving.

Treatment

The mainstay of treatment is the assessment of dehydration and replacement of fluid and electrolytes. Administration of antibiotics has been shown to shorten the course of illness and duration of excretion of enterotoxigenic E. coli (ETEC) in adults in endemic areas and in traveller's diarrhea, though the rate of resistance to commonly used antibiotics is increasing and they are generally not recommended. The antibiotic used depends upon susceptibility patterns in the particular geographical region. Currently, the antibiotics of choice are fluoroquinolones or azithromycin, with an emerging role for rifaximin. Oral rifaximin, a semisynthetic rifamycin derivative, is an effective and well-tolerated antibacterial for the management of adults with non-invasive traveller's diarrhea. Rifaximin was significantly more effective than placebo and no less effective than ciprofloxacin in reducing the duration of diarrhea. While rifaximin is effective in patients with E. coli-predominant traveller's diarrhea, it appears ineffective in patients infected with inflammatory or invasive enteropathogens.

Prevention

ETEC is the type of E. coli that most vaccine development efforts are focused on. Antibodies against the LT and major CFs of ETEC provide protection against LT-producing, ETEC-expressing homologous CFs. Oral inactivated vaccines consisting of toxin antigen and whole cells, i.e. the licensed recombinant cholera B subunit (rCTB)-WC cholera vaccine Dukoral, have been developed. There are currently no licensed vaccines for ETEC, though several are in various stages of development. In different trials, the rCTB-WC cholera vaccine provided high (85–100%) short-term protection. An oral ETEC vaccine candidate consisting of rCTB and formalin inactivated E. coli bacteria expressing major CFs has been shown in clinical trials to be safe, immunogenic, and effective against severe diarrhoea in American travelers but not against ETEC diarrhoea in young children in Egypt. A modified ETEC vaccine consisting of recombinant E. coli strains over-expressing the major CFs and a more LT-like hybrid toxoid called LCTBA, are undergoing clinical testing.

Other proven prevention methods for E. coli transmission include handwashing and improved sanitation and drinking water, as transmission occurs through fecal contamination of food and water supplies. Additionally, thoroughly cooking meat and avoiding consumption of raw, unpasteurized beverages, such as juices and milk are other proven methods for preventing E.coli. Lastly, avoid cross-contamination of utensils and work spaces when preparing food.

Model Organism in Life Science Research

Because of its long history of laboratory culture and ease of manipulation, E. coli plays an important role in modern biological engineering and industrial microbiology. The work of Stanley Norman Cohen and Herbert Boyer in E. coli, using plasmids and restriction enzymes to create recombinant DNA, became a foundation of biotechnology.

E. coli is a very versatile host for the production of heterologous proteins, and various protein expression systems have been developed which allow the production of recombinant proteins in E. coli. Researchers can introduce genes into the microbes using plasmids which permit high level expression of protein, and such protein may be mass-produced in industrial fermentation processes. One of the first useful applications of recombinant DNA technology was the manipulation of E. coli to produce human insulin.

Many proteins previously thought difficult or impossible to be expressed in E. coli in folded form have been successfully expressed in E. coli. For example, proteins with multiple disulphide bonds may be produced in the periplasmic space or in the cytoplasm of mutants rendered sufficiently oxidizing to allow disulphide-bonds to form, while proteins requiring post-translational modification such as glycosylation for stability or function have been expressed using the N-linked glycosylation system of Campylobacter jejuni engineered into E. coli. Modified E. coli cells have been used in vaccine development, bioremediation, production of biofuels, lighting, and production of immobilised enzymes.

Strain K-12 is a mutant form of E. coli that over-expresses the enzyme Alkaline Phosphatase (ALP). The mutation arises due to a defect in the gene that constantly codes for the enzyme. A gene that is producing a product without any inhibition is said to have constitutive activity. This particular mutant form is used to isolate and purify the aforementioned enzyme. Strain OP50 of Escherichia coli is used for maintenance of Caenorhabditis elegans cultures.

Strain JM109 is a mutant form of E. coli that is recA and endA deficient. The strain can be utilized for blue/white screening when the cells carry the fertility factor episome Lack of recA decreases the possibility of unwanted restriction of the DNA of interest and lack of endA inhibit plasmid DNA decomposition. Thus, JM109 is useful for cloning and expression systems.

Model Organism

E. coli is frequently used as a model organism in microbiology studies. Cultivated strains (e.g. E. coli K12) are well-adapted to the laboratory environment, and, unlike wild-type strains, have lost their ability to thrive in the intestine. Many laboratory strains lose their ability to form biofilms. These features protect wild-type strains from antibodies and other chemical attacks, but require a large expenditure of energy and material resources. E. coli is often used as a representative microorganism in the research of novel water treatment and sterilisation methods, including photocatalysis. By standard plate count methods, following sequential dilutions, and growth on agar gel plates, the concentration of viable organisms or CFUs (Colony Forming Units), in a known volume of treated water can be evaluated, allowing the comparative assessment of materials performance.

In 1946, Joshua Lederberg and Edward Tatum first described the phenomenon known as bacterial conjugation using E. coli as a model bacterium, and it remains the primary model to study conjugation. E. coli was an integral part of the first experiments to understand phage genetics, and early researchers, such as Seymour Benzer, used E. coli and phage T4 to understand the topography of gene structure. Prior to Benzer's research, it was not known whether the gene was a linear structure, or if it had a branching pattern.

From 2002 to 2010, a team at the Hungarian Academy of Science created a strain of Escherichia coli called MDS42, which is now sold by Scarab Genomics of Madison, WI under the name of "Clean Genome. E.coli", where 15% of the genome of the parental strain (E. coli K-12 MG1655) were removed to aid in molecular biology efficiency, removing IS elements, pseudogenes and phages, resulting in better maintenance of plasmid-encoded toxic genes, which are often inactivated by transposons. Biochemistry and replication machinery were not altered.

By evaluating the possible combination of nanotechnologies with landscape ecology, complex habitat landscapes can be generated with details at the nanoscale. On such synthetic ecosystems, evolutionary experiments with E. coli have been performed to study the spatial biophysics of adaptation in an island biogeography on-chip.

Studies are also being performed attempting to program E. coli to solve complicated mathematics problems, such as the Hamiltonian path problem.

PURPLE BACTERIA

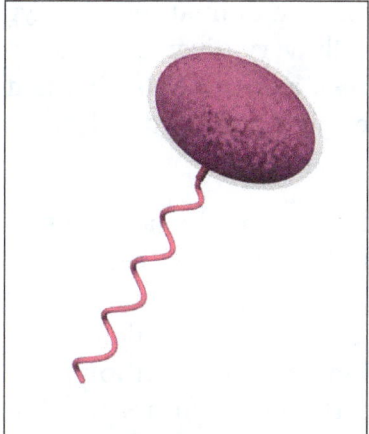

Rhodobacter spheroides with its single flagellum inserted in the side of the cell.

The purple bacteria are one of the groups of photosynthetic bacteria (the other main groups being the cyanobacteria and green bacteria, though halobacterium can use light as an energy source it is not photosynthetic). They fall into two groups, according to their metabolism (each group appears to be an assembly of different lineages that evolved separately): purple non-sulphur bacteria (e.g. Rhodospirillum, Rhodobacter) and purple sulphur bacteria. Purple sulphur bacteria can use elemental sulphur and also sulphide (e.g. hydrogen sulphide) as an electron-donor in respiration, whilst non-sulphur bacteria can not, but use an organic electron-donor instead (e.g. succinate, malate) or elemental hydrogen.

Purple Sulphur and Purple Non-sulphur Bacteria and Bacterial Energy Metabolism

Bacterial metabolism is a very complex topic, because bacteria are very metabolically diverse. The purple bacteria are good examples to illustrate many of the features.

Respiration is a complex redox (reduction-oxidation) reaction (consisting of a chain of smaller redox reactions) in which fuel (the initial electron donor) is oxidised to release energy, that is the fuel loses one or more electrons, which carry the energy, by passing them onto the first link in an electron transport chain (ETC, a chain of specific biochemicals) which gains the electron and so becomes reduced (OIL RIG: oxidation is loss of electrons, reduction is gain). The electron is then passed sequentially from link-to-link down the chain, as components of the chain become alternately oxidised and reduced whilst the electron loses energy at each step (this flow of electrons is essentially the flow of electric charge or electricity). Much of the energy lost by the electron is harvested and converted into chemical energy where it is stored until needed, usually in molecules of ATP. (Initially the energy is used to pump protons across the (inner) cell membrane and these protons then flow through the ATPase molecular motors, causing them to rotate and this rotaional mechanical energy is converted into chemical energy in ATP). The electron finally exits the ETC, having lost much of its energy, and reduces the final electron-acceptor, which in animal cells is oxygen which become reduced to water. Bacteria are metabolically very diverse and some make use of other electron acceptors (reducible substances), according to what is most readily available in their habitat. Similarly they can use a variety of initial electron donors (oxidisable substances or fuels).

Photosynthesis is also a redox chain which harnesses light energy to build complex organic molecules (including fuels for respiration) from simple carbon compounds, including carbon dioxide (as in plants) or simple water-soluble organic carbon sources like acetate and other organic acids. Photosynthesis must always be accompanied by respiration - it is an addition not an alternative, it simply uses light energy to generate the carbon fuels for respiration (such as sugars like maltose and glucose) and other organic building blocks (like amino acids, lipids and nucleotides). Light oxidises the light-receptive pigments (chlorophylls and carotenoids in plants, bacteriochlorophyll in bacteria) which pass an electron through an ETC. The electron loses energy as it passes down the ETC and this energy is used to build organic molecules from the simpler carbon sources (like carbon dioxide or acetate). In non-cyclic systems, the electron must arrive at a final electron-acceptor (NADP in plants). In cyclic systems, there is no net oxidation or reduction, rather electrons are channeled through a circuit. An electron donor is needed to replace the electron lost by the chlorophyll in plants and this is water which is split by plants, generating oxygen.

Heterotrophs are organisms, like animals, which obtain their carbon by breaking-down organic molecules built by other organisms. The term usually refers to chemoheterotrophs, like animals, who use these ready-made carbon-sources also as fuels, to supply energy by respiration. However some bacteria are phototoheterotrophs, using light as a source of energy (rather than chemical fuel) but ready-made organic sources for a supply of carbon.

Autotrophs make their own organic molecules from simple carbon molecules readily available in the environment such as carbon dioxide from the air (we can think of these materials as readily available abiotically, though in reality natural cycles involving living organisms are needed to maintain supplies). In the case of utilising a gas, like carbon dioxide, and using it to make 'solid' materials like proteins, we say that the carbon has been fixed (made 'solid'). We usually think of plants which are photoautotrophs, using light as an energy source and carbon dioxide as a carbon source to build their own materials by photosynthesis. However, some bacteria are chemoautotrophs, using chemical fuels as a source of energy to synthesise their own materials from simple carbon sources.

Chemotrophs is a generic term referring to organisms that use chemical reactions as their chief source of energy, that is chemoheterotrophs and chemoautotrophs collectively. (Lit. 'chemical-eating').

Phototrophs similar refers to photoautotrophs and photoheterotrophs - organisms that use light as thei chief source of energy. (Lit. 'light-eating').

- Purple bacteria can grow photoautotrophically (photosynthetically using light to fix a simple inorganic carbon source) in anaerobic conditions (in the absence of air) using carbon dioxide as the carbon-source and hydrogen as the electron donor.

- They can also grow photoheterotrophically (photosynthetically using light and an organic carbon source) under anaerobic conditions, on such organic carbon sources as acetate, malate and succinate (salts of organic acids) which act as electron donors.

- Some can also grow as chemoheterotrophs (heterotrophs, by respiring organic fuels/carbon sources) in the dark, under aerobic conditions (using oxygen as the final electron acceptor). Some non-sulphur types only.

Photosynthesis

In bacteria, as in plants, antenna pigments trap light energy and then pass this on to light-sensitive pigments in the reaction centres (which may also catch light directly, but the antenna pigments greatly enhance the efficiency of photon capture). Purple bacteria contain bacteriochlorophyll type a or b as the primary light-sensitive pigment in the reaction centres and antennae, and in the antenna carotenoids enhance the range of wavelengths of light that can be harnessed. These carotenoids may be purple, red, brown or orange in colour and give the bacteria their characteristic colouration.

Purple bacteria are Gram negatives and so possess a double membrane in the cell envelope. The inner membrane is extensively folded into sacs, tubes or sheets, increasing its surface area, and most of the photosynthetic machinery is located on these folds. These folds contain 3 types of molecular complexes involved in photosynthesis:

- Antenna or light-harvesting complexes (LHC);

- Reaction centres (RC);

- Cytochrome (cyt) bc1 complex.

A biomolecular complex is a close association of biochemicals into a working unit, which minimises the distances reactants must diffuse in order to react to one-another and greatly increases efficiency (a ribosome is another example of a biomolecular complex). The antenna or LHC binds carotenoids and bacteriochlorophyll (bchl) molecules and collects incident light. In most purple bacteria, including Rhodobacter sphaeroides, there are two types: LHC1 and LHC2. The amount of LHC2 is variable and dependent on environmental conditions, such as the amount of oxygen present and light intensity. The ratio of LHC1 to RC is fixed as these combine to form a RC-LHC1 supercomplex. There are as many as 100 bacteriochlorophyll molecules in each RC, increasing the amount of incident light that can be absorbed. At the core of each RC is a bchl-dimer (bacteriochlorophyll dimer, two bchl molecules joined together).

Photosynthesis in these organisms is cyclic, and does not produce oxygen. The sequence of events, as worked-out for Rhodobacter sphaeroides, is as follows:

- A photon is absorbed by the LHC, causing it to lose an electron to the RC in less than 100 ps (ps = picosecond). If absorbed by LHC2 then the electron is passed to LHC1.

- The electron is passed to bchl in the RC and on to the core bchl-dimer in the RC which acts as the primary electron donor for the ETC and passes an electron to bacteriophytin in 2-3 ps.

- Bacteriophytin passes an electron to the next link in the cycle, a quinone molecule (QA) in about 200 ps.

- The electron is passed on to a second quinone (QB).

- Steps 1-4 are repeated until QB is carrying two electrons.

- The reduced QB picks up two protons from the cytosol, forming QBH_2 (quinol).

The quinol is released from the RC and delivers its two protons to the periplasm (the region between the inner and outer cell membranes), a reaction mediated by the Rieske protein (which contains catalytic Fe_2S_2 iron sulphide clusters) and the cyt bc1 complex. The electrons are passed on to cytochrome c_2 (cyt c_2).

Cyt c_2 completes the cycle by passing the electrons back to bacteriochlorophyll, replacing the electrons lost initially.

The tremendous speed with which the electron is initially transferred from the LHC to the RC ensures very efficient electron capture. The quantum yield (ratio of electrons produced per photon absorbed) is almost 1. (There is always a possibility that a molecule will absorb a photon and de-excite by some other means, with the energy wasted. This can never be eliminated, but it can be minimised).

Under certain conditions, Rhodobacter sphaeroides produces membrane invaginations free of LHC_2, this simplified membrane-system has been studied in great structural detail and it is found to be studded in complexes as shown below.

In the figure, Left: the membrane supercomplexes of Rhodobacter each repeating unit is 19.8 by 11.2 nanometres. In vitro, two rings of LHC1 form around the RC, however, in vivo, each ring is open, resulting in an S-shaped supercomplex, likely due to LHC1. The regions inside each incomplete ring of the S-upercomplex (shown stipled) is likely to be a RC. The small detached structure between adjacent S-supercomplexes could be the cytochrome bc1 complex and then quinones would have only a short distance to diffuse bewteen the RC and the cyt bc1 (diffusing through the gaps in the S-shaped rings).

The protein PufX is needed for the formation of S-shaped dimers, as opposed to closed rings, and this protein may thus function to prevent formation of a closed ring, allowing quinones to diffuse to the cyt bc1 and back.

The grouping of the photosystem into these highly organised membrane complexes and supercomplexes sppeds up the reaction, since diffusion is kept to a minimum. Indeed, the system still functions in membranes frozen at -20 C, illustrated how the system is not diffusion limited. This is an excellent example of how the nanomachinery of cells can be highly organised, something which becomes ever increasingly apparent with ongoing research. The old model of a cell as a 'bag of chemicals' is far from the truth.

Locomotion and Chemokinesis/Chemotaxis

Rhodobacter sphaeroides demonstrates an interesting variation on bacterial locomotion, which is quite different from the run and tumble behaviour of peritrichously flagellated Escherichia coli. Rhodobacter will undergo chemokinesis towards nutrients, that is positive chemokinesis. This is a form of klinokinesis. In klinokinesis changes in turning frequency occur in response to changes in stimulus strength. Turning frequency changes in response to temporal changes in stimulant concentration. As in Escherichia coli, the cells swim about and so compare stimulus strengths at different spatial positions at different times. They have a cellular memory (the response habituates) and so are able to detect a drop or increase in stimulant concentration as they move about and hence indirectly detect chemical gradients. This is not chemotaxis, as defined in ethology, in which the gradient is measured directly in one instant of time at each spatial position, for example by having widely spaced sensors to give a stereo view of the odour gradient. Only the very largest bacteria exhibit true chemotaxis, most are simply too small to measure spatial gradients directly.

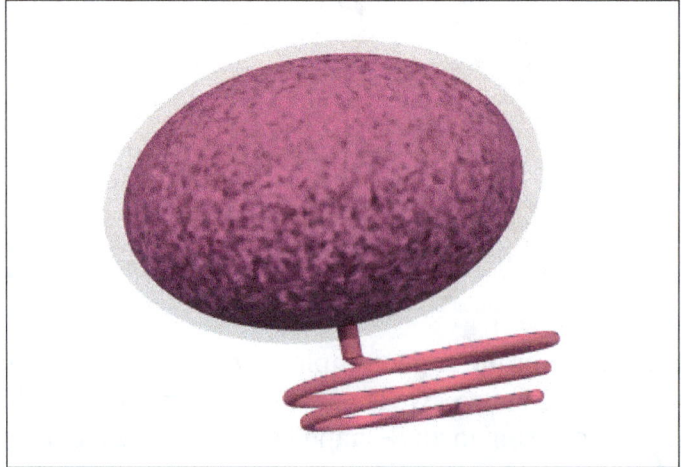

Having only one flagellum necessitates a different mechanism of cell turning than the random tumbling of Escherichia coli. Some bacteria with a single polar flagellum alternate the direction of flagellum rotation, causing it to push or pull. Similarly, bacteria with one or more flagella at each pole can switch each bundle from pushing to pulling. However, the flagellum of Rhodobacter does not change direction, it can only rotate clockwise (or counterclockwise ion some strains) and is a right-handed helix. However, it can coil-up close to the cell body (by reducing wavelength and increasing amplitude, the switch occuring from the tip towards the base) and rotate slowly. This slowly rotating coil turns the cell (in conjunction with Brownian motion) and is functionally similar to the tumbles exhibited by Escherichia coli. This is the mode of chemokinesis demonstrated under anaerobic conditions. Positive chemokinesis in response to nutrients has been observed.

The response of Rhodobacter is pessimistic: it increases turning frequency as it moves down an attractant gradient. This is in contrast to the optimistic response shown by Escherichia coli in which tumbling frequency reduces as it moves up an attractant gradient. Rhodobacter swims at about 35 micrometres per second (compared to 20-30 micrometres per second for Escherichia coli).

In aerobic conditions, the behaviour becomes more complicated and changes in flagella rotation speed can also cause the cell to turn in an arc when the rotation rate slows. A change in speed of locomotion in response to a change in strength of a stimulus is orthokinesis (though in microbiology literature this is usually erroneously defined exclusively as chemokinesis, but chemokinesis can involve changes in turning frequency and speed of locomotion - the terms 'kinesis' and 'kinetic' refer to motion or velocity, that is speed and direction, so both orthokinesis and klinokinesis are types of kinesis).

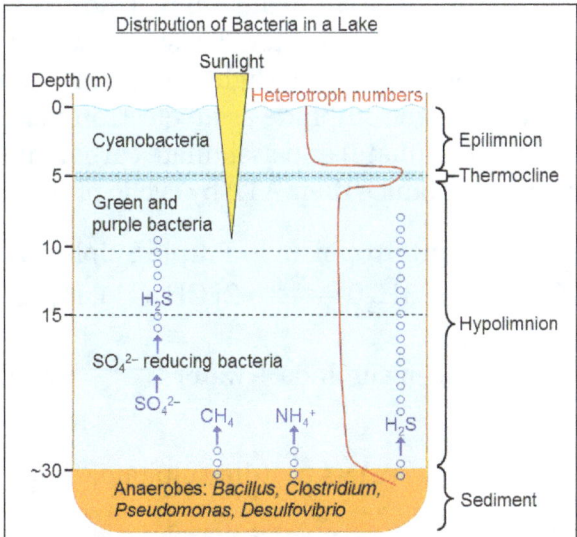

The diagram above shows the distribution of bacteria in a typical freshwater lake in which the water is stratified. The thermocline is the depth at which the temperature suddenly changes and defines the epilimnion above and the hypolimnion below. In the epilmnion, photosynthetic bacteria, such as the cyanobacteria, may contribute significantly to primary productivity (the fixation of carbon dioxide into organic carbon, in this case by photosynthesis requiring light). (Eukaryotic algae usually make the larger contribution). If the lake is more-or-less stagnant, then the sediments will be anoxic (lacking in oxygen) and anaerobic bacteria will dominate here. These bacteria may be strict anaerobes, that can only grow in the absence of oxygen (like Clostridium) or organisms that can grow anaerobically when they need to (like Pseudomonas). Heterotrophs are those bacteria that break-down organic materials as a fuel and carbon-source. These are most abundant in the thermocline (aerobes) and in the sediments (anaerobes) or wherever organic material is most abundant. Sulphate may be injected into the lake in sea water or sulphate-rich springs.

The products of anaerobic decomposition of these organic materials (e.g. the remains of dead organisms) include methane gas (CH_4), ammonia gas (NH_3) which dissolves to give ammonium (NH_4^+) and hydrogen sulphide gas (H_2S). Hydrogen sulphide gas is toxic and if too much is produced, as it can be in stagnant waters with excess organic materials, then fish and other organisms may die. It is fortunate. Therefore, that higher in the water column there are the green and purple sulphur bacteria which can utilise this hydrogen sulphide and convert it into elemental sulphur. These bacteria require some light and so occur near the surface, below the cyanobacteria, where there is still sufficient light, but where their sulphide source can be found. These organisms keep the upper oxygenated waters largely free of hydrogen sulphide in a healthy lake.

Purple Sulphur Bacteria

E.g. Thiospirillum, Ectothiorhodospira, Chromatium, Thiocystis, Thiocapsa, Lamprocystis, Thiodictyon, Thiopedia, Amoebobacter. These are strictly anaerobic and grow predominantly photoautotrophically (using light to fix carbon and hydrogen as the electron donor). They use hydrogen sulphide as an electron donor (oxidising it to sulphur and then more slowly to sulphate) and assimilate carbon dioxide (largely through the Calvin-Benson cycle) and produce their ATP by cyclic photophosphorylation.

Anaerobic Photoautotrophism in Purple Sulphur Bacteria;

$$2CO_2 + H_2 + 3H_2O \xrightarrow{\text{light}} 2(CH_2O) + H_2SO_4$$

Carbon dioxide + hydrogen sulphide + water $\xrightarrow{\text{light}}$ carbohydrate + sulphuric acid

(The sulphuric acid reacts to from sulphate, SO_2^{4-}, salts).

Results:

- CO_2 is assimilated into organic carbon

- H_2S is oxidised to S, stored as granules inside the cytoplasm (extracellularly in Ectothiorhodospira), which is more slowly oxidised to SO_4^{2-}

Purple sulphur bacteria can also grow photoheterotrophically, that is using light and an organic carbon source.

Anaerobic Photoheterotropic Growth in Purple Sulphur Bacteria

- Carbon is assimilated not from carbon dioxide, but from small organic acids/salts such as acetate (=ethanoate, CH_3COO-) as carbon sources and electron donors, requires light;

- Some require a small amount of H_2S for this process, as a sulphur source (e.g. for making proteins) since they can not use sulphate.

Purple sulphur bacteria are obligate phototrophs, and require light in order to grow.

Purple Non-sulphur Bacteria

E.g. Rhodospirillum, Rhodopseudomonas, Rhodomicrobium, Rhodopila, Rhodocyclus, Rhodobacter.

These are predominantly anaerobic photoheterotrophic using fatty acids, other organic acids, primary and secondary alcohols, carbohydrates and aromatic organic compounds as their carbon source. They can also respire these compounds in the dark (chemoheterotrophism).

They are sensitive to hydrogen sulphide, which inhibits their growth, although some can oxidise very low levels of sulphide anaerobically in the light. They typically occur in freshwater lakes and ponds where organic matter is present but sulphide is low.

Rhodobacter capsulatus is able to grow as an aerobic chemoautotroph (assimilating carbon but by using chemical energy rather than light) in the dark with hydrogen as the electron donor. Contrast this with purple sulphur bacteria which are obligate phototrophs.

Denitrification

Some purple non-sulphur bacteria, e.g. Rhodopseudomonas palustris and Rhodobacter sphaeroides, can denitrify, that is reduce nitrate (which is reduced to nitrite then to nitrogen, with each denitrifying bacterium capable of one or both stages) by respiring it anaerobically instead of oxygen when oxygen is exhausted.

GRAM POSITIVE BACTERIA

Prokaryotes are identified as gram-positive if they have a multiple layer matrix of peptidoglycan forming the cell wall. Crystal violet, the primary stain of the Gram stain procedure, is readily retained and stabilized within this matrix, causing gram-positive prokaryotes to appear purple under a brightfield microscope after Gram staining. For many years, the retention of Gram stain was one of the main criteria used to classify prokaryotes, even though some prokaryotes did not readily stain with either the primary or secondary stains used in the Gram stain procedure.

Advances in nucleic acid biochemistry have revealed additional characteristics that can be used to classify gram-positive prokaryotes, namely the guanine to cytosine ratios (G+C) in DNA and the composition of 16S rRNA subunits. Microbiologists currently recognize two distinct groups of gram-positive, or weakly staining gram-positive, prokaryotes. The class Actinobacteria comprises the high G+C gram-positive bacteria, which have more than 50% guanine and cytosine nucleotides in their DNA. The class Bacilli comprises low G+C gram-positive bacteria, which have less than 50% of guanine and cytosine nucleotides in their DNA.

Actinobacteria: High G+C Gram positive Bacteria

Their microscopic appearance can range from thin filamentous branching rods to coccobacilli. Some Actinobacteria are very large and complex, whereas others are among the smallest independently living organisms. Most Actinobacteria live in the soil, but some are aquatic. The vast majority are aerobic. One distinctive feature of this group is the presence of several different peptidoglycans in the cell wall.

The genus Actinomyces is a much studied representative of Actinobacteria. Actinomyces spp. play an important role in soil ecology, and some species are human pathogens. A number of Actinomyces spp. inhabit the human mouth and are opportunistic pathogens, causing infectious diseases like periodontitis (inflammation of the gums) and oral abscesses. The species A. israelii is an anaerobe notorious for causing endocarditis (inflammation of the inner lining of the heart).

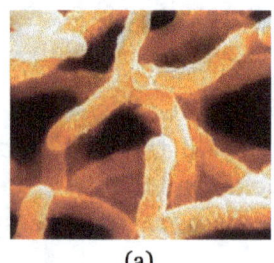
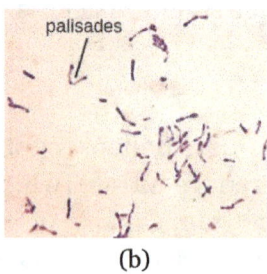
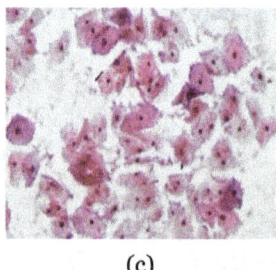

(a)   (b)   (c)

In figure, (a) Actinomyces israelii (false-color scanning electron micrograph [SEM]) has a branched structure. (b) Corynebacterium diphtheria causes the deadly disease diphtheria. Note the distinctive palisades. (c) The gram-variable bacterium Gardnerella vaginalis causes bacterial vaginosis in women. This micrograph shows a Pap smear from a woman with vaginosis.

The genus Mycobacterium is represented by bacilli covered with a mycolic acid coat. This waxy coat protects the bacteria from some antibiotics, prevents them from drying out, and blocks penetration by Gram stain reagents. Because of this, a special acid-fast staining procedure is used to visualize these bacteria. The genus Mycobacterium is an important cause of a diverse group of infectious diseases. M. tuberculosis is the causative agent of tuberculosis, a disease that primarily impacts the lungs but can infect other parts of the body as well. It has been estimated that one-third of the world's population has been infected with M. tuberculosis and millions of new infections occur each year. Treatment of M. tuberculosis is challenging and requires patients to take a combination of drugs for an extended time. Complicating treatment even further is the development and spread of multidrug-resistant strains of this pathogen.

Another pathogenic species, M. leprae, is the cause of Hansen's disease (leprosy), a chronic disease that impacts peripheral nerves and the integrity of the skin and mucosal surface of the respiratory tract. Loss of pain sensation and the presence of skin lesions increase susceptibility to secondary injuries and infections with other pathogens.

Bacteria in the genus Corynebacterium contain diaminopimelic acid in their cell walls, and microscopically often form palisades, or pairs of rod-shaped cells resembling the letter V. Cells may contain metachromatic granules, intracellular storage of inorganic phosphates that are useful for identification of Corynebacterium. The vast majority of Corynebacterium spp. are nonpathogenic; however, C. diphtheria is the causative agent of diphtheria, a disease that can be fatal, especially in children. C. diphtheria produces a toxin that forms a pseudomembrane in the patient's throat, causing swelling, difficulty breathing, and other symptoms that can become serious if untreated.

The genus Bifidobacterium consists of filamentous anaerobes, many of which are commonly found in the gastrointestinal tract, vagina, and mouth. In fact, Bifidobacterium spp. constitute a substantial part of the human gut microbiota and are frequently used as probiotics and in yogurt production.

The genus Gardnerella, contains only one species, G. vaginalis. This species is defined as "gram-variable" because its small coccobacilli do not show consistent results when Gram stained. Based on its genome, it is placed into the high G+C gram-positive group. G. vaginalis can cause bacterial vaginosis in women; symptoms are typically mild or even undetectable, but can lead to complications during pregnancy.

Table: Summarizes the characteristics of some important genera of Actinobacteria.

Actinobacteria: High G+C Gram-Positive		
Example Genus	Microscopic Morphology	Unique Characteristics
Actinomyces	Gram-positive bacillus; in colonies, shows fungus-like threads (hyphae)	Facultative anaerobes; in soil, decompose organic matter; in the human mouth, may cause gum disease.
Arthrobacter	Gram-positive bacillus (at the exponential stage of growth) or coccus (in stationary phase)	Obligate aerobes; divide by "snapping," forming V-like pairs of daughter cells; degrade phenol, can be used in bioremediation.

Bifidobacterium	Gram-positive, filamentous actinobacterium	Anaerobes commonly found in human gut microbiota.
Corynebacterium	Gram-positive bacillus	Aerobes or facultative anaerobes; form palisades; grow slowly; require enriched media in culture; C. diphtheriae causes diphtheria.
Frankia	Gram-positive, fungus-like (filamentous) bacillus	Nitrogen-fixing bacteria; live in symbiosis with legumes.
Gardnerella	Gram-variable coccobacillus	Colonize the human vagina, may alter the microbial ecology, thus leading to vaginosis.
Micrococcus	Gram-positive coccus, form microscopic clusters	Ubiquitous in the environment and on the human skin; oxidase-positive (as opposed to morphologically similar S. aureus); some are opportunistic pathogens.
Mycobacterium	Gram-positive, acid-fast bacillus	Slow growing, aerobic, resistant to drying and phagocytosis; covered with a waxy coat made of mycolic acid; M. tuberculosis causes tuberculosis; M. leprae causes leprosy.
Nocardia	Weakly gram-positive bacillus; forms acid-fast branches	May colonize the human gingiva; may cause severe pneumonia and inflammation of the skin.
Propionibacterium	Gram-positive bacillus	Aerotolerant anaerobe; slow-growing; P. acnes reproduces in the human sebaceous glands and may cause or contribute to acne.
Rhodococcus	Gram-positive bacillus	Strict aerobe; used in industry for biodegradation of pollutants; R. fascians is a plant pathogen, and R. equi causes pneumonia in foals.
Streptomyces	Gram-positive, fungus-like (filamentous) bacillus	Very diverse genus (>500 species); aerobic, spore-forming bacteria; scavengers, decomposers found in soil (give the soil its "earthy" odor); used in pharmaceutical industry as antibiotic producers (more than two-thirds of clinically useful antibiotics)

Low G+C Gram positive Bacteria

The low G+C gram-positive bacteria have less than 50% guanine and cytosine in their DNA, and this group of bacteria includes a number of genera of bacteria that are pathogenic.

Clostridia

One large and diverse class of low G+C gram-positive bacteria is Clostridia. The best studied genus of this class is Clostridium. These rod-shaped bacteria are generally obligate anaerobes that produce endospores and can be found in anaerobic habitats like soil and aquatic sediments rich in organic nutrients. The endospores may survive for many years.

Clostridium spp. produce more kinds of protein toxins than any other bacterial genus, and several species are human pathogens. C. perfringens is the third most common cause of food poisoning in the United States and is the causative agent of an even more serious disease called gas gangrene. Gas gangrene occurs when C. perfringens endospores enter a wound and germinate, becoming viable bacterial cells and producing a toxin that can cause the necrosis (death) of tissue. C. tetani, which causes tetanus, produces a neurotoxin that is able to enter neurons, travel to regions of the central nervous system where it blocks the inhibition of nerve impulses involved in muscle contractions, and cause a life-threatening spastic paralysis. C. botulinum produces botulinum neurotoxin, the most lethal biological toxin known. Botulinum toxin is responsible for rare but frequently fatal cases of botulism. The toxin blocks the release of acetylcholine in neuromuscular junctions, causing flaccid paralysis. In very small concentrations, botulinum toxin has been used to treat muscle pathologies in humans and in a cosmetic procedure to eliminate wrinkles. C. difficile is a common source of hospital-acquired infections that can result in serious and even fatal cases of colitis (inflammation of the large intestine). Infections often occur in patients who are immunosuppressed or undergoing antibiotic therapy that alters the normal microbiota of the gastrointestinal tract.

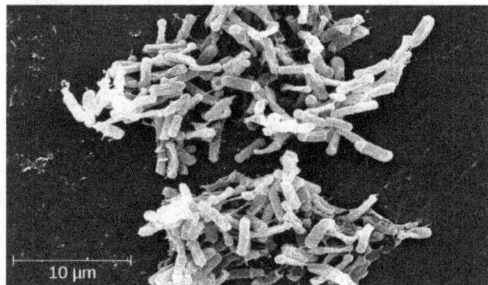

Clostridium difficile, a gram-positive, rod-shaped bacterium, causes severe colitis and diarrhea, often after the normal gut microbiota is eradicated by antibiotics.

Lactobacillales

The order Lactobacillales comprises low G+C gram-positive bacteria that include both bacilli and cocci in the genera Lactobacillus, Leuconostoc, Enterococcus, and Streptococcus. Bacteria of the latter three genera typically are spherical or ovoid and often form chains.

Streptococcus, is responsible for many types of infectious diseases in humans. Species from this genus, often referred to as streptococci, are usually classified by serotypes called Lancefield groups, and by their ability to lyse red blood cells when grown on blood agar.

S. pyogenes belongs to the Lancefield group A, β-hemolytic Streptococcus. This species is considered a pyogenic pathogen because of the associated pus production observed with infections it causes. S. pyogenes is the most common cause of bacterial pharyngitis (strep throat); it is also an important cause of various skin infections that can be relatively mild (e.g., impetigo) or life threatening (e.g., necrotizing fasciitis, also known as flesh eating disease), life threatening.

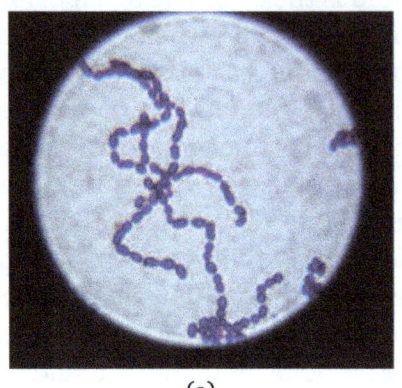 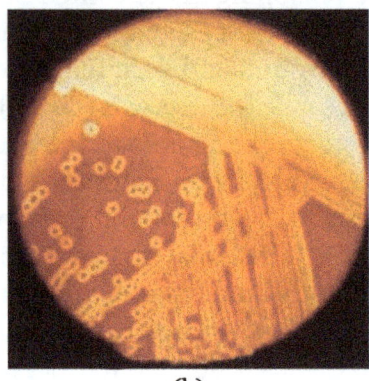

(a)   (b)

(a) A gram-stained specimen of Streptococcus pyogenes shows the chains of cocci characteristic of this organism's morphology. (b) S. pyogenes on blood agar shows characteristic lysis of red blood cells, indicated by the halo of clearing around colonies.

The nonpyogenic (i.e., not associated with pus production) streptococci are a group of streptococcal species that are not a taxon but are grouped together because they inhabit the human mouth. The nonpyogenic streptococci do not belong to any of the Lancefield groups. Most are commensals, but a few, such as S. mutans, are implicated in the development of dental caries.

S. pneumoniae (commonly referred to as pneumococcus), is a Streptococcus species that also does not belong to any Lancefield group. S. pneumoniae cells appear microscopically as diplococci, pairs of cells, rather than the long chains typical of most streptococci. Scientists have known since the 19th century that S. pneumoniae causes pneumonia and other respiratory infections. However, this bacterium can also cause a wide range of other diseases, including meningitis, septicemia, osteomyelitis, and endocarditis, especially in newborns, the elderly, and patients with immunodeficiency.

Bacilli

The name of the class Bacilli suggests that it is made up of bacteria that are bacillus in shape, but it is a morphologically diverse class that includes bacillus-shaped and coccus-shaped genera. Among the many genera in this class are two that are very important clinically: Bacillus and Staphylococcus.

Bacteria in the genus Bacillus are bacillus in shape and can produce endospores. They include aerobes or facultative anaerobes. A number of Bacillus spp. are used in various industries, including the production of antibiotics (e.g., barnase), enzymes (e.g., alpha-amylase, BamH1 restriction endonuclease), and detergents (e.g., subtilisin).

Two notable pathogens belong to the genus Bacillus. B. anthracis is the pathogen that causes anthrax, a severe disease that affects wild and domesticated animals and can spread from infected animals to humans. Anthrax manifests in humans as charcoal-black ulcers on the skin, severe enterocolitis, pneumonia, and brain damage due to swelling. If untreated, anthrax is lethal. B. cereus, a closely related species, is a pathogen that may cause food poisoning. It is a rod-shaped species that forms chains. Colonies appear milky white with irregular shapes when cultured on blood agar. One other important species is B. thuringiensis. This bacterium produces a number of substances used as insecticides because they are toxic for insects.

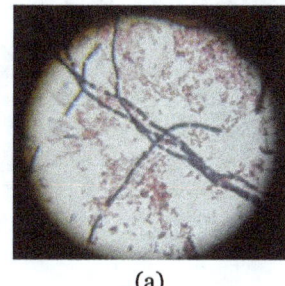 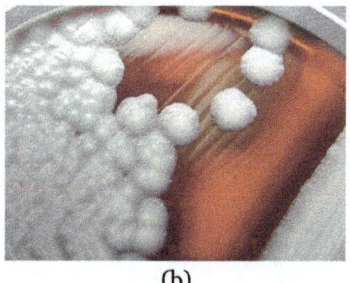

(a)   (b)

(a) In this gram-stained specimen, the violet rod-shaped cells forming chains are the gram-positive bacteria Bacillus cereus. The small, pink cells are the gram-negative bacteria Escherichia coli.
(b) In this culture, white colonies of B. cereus have been grown on sheep blood agar.

The genus Staphylococcus also belongs to the class Bacilli, even though its shape is coccus rather than a bacillus. Staphylococcus spp. are facultative anaerobic, halophilic, and nonmotile. The two best-studied species of this genus are S. epidermidis and S. aureus.

S. epidermidis, whose main habitat is the human skin, is thought to be nonpathogenic for humans with healthy immune systems, but in patients with immunodeficiency, it may cause infections in skin wounds and prostheses (e.g., artificial joints, heart valves). S. epidermidis is also an important cause of infections associated with intravenous catheters. This makes it a dangerous pathogen in hospital settings, where many patients may be immunocompromised.

Strains of S. aureus cause a wide variety of infections in humans, including skin infections that produce boils, carbuncles, cellulitis, or impetigo. Certain strains of S. aureus produce a substance called enterotoxin, which can cause severe enteritis, often called staph food poisoning. Some strains of S. aureus produce the toxin responsible for toxic shock syndrome, which can result in cardiovascular collapse and death.

Many strains of S. aureus have developed resistance to antibiotics. Some antibiotic-resistant strains are designated as methicillin-resistant S. aureus (MRSA) and vancomycin-resistant S. aureus (VRSA). These strains are some of the most difficult to treat because they exhibit resistance to nearly all available antibiotics, not just methi-cillin and vancomycin. Because they are difficult to treat with antibiotics, infections can be lethal. MRSA and VRSA are also contagious, posing a serious threat in hospitals, nursing homes, dialysis facilities, and other places where there are large populations of elderly, bedridden, and immunocompromised patients.

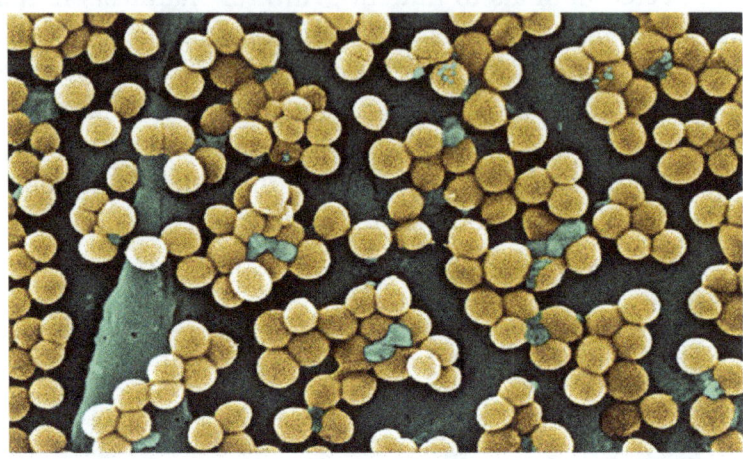

This SEM of Staphylococcus aureus illustrates the typical "grape-like" clustering of cells.

Mycoplasmas

Although Mycoplasma spp. do not possess a cell wall and, therefore, are not stained by Gram-stain reagents, this genus is still included with the low G+C gram-positive bacteria. The genus Mycoplasma includes more than 100 species, which share several unique characteristics. They are very small cells, some with a diameter of about 0.2 μm, which is smaller than some large viruses. They have no cell walls and, therefore, are pleomorphic, meaning that they may take on a variety of shapes and can even resemble very small animal cells. Because they lack a characteristic shape, they can be difficult to identify. One species, M. pneumoniae, causes the mild form of pneumonia known as "walking pneumonia" or "atypical pneumonia." This form of pneumonia is typically less severe than forms caused by other bacteria or viruses.

Table: Summarizes the characteristics of notable genera low G+C Gram-positive bacteria.

Bacilli: Low G+C Gram-Positive Bacteria		
Example Genus	Microscopic Morphology	Unique Characteristics
Bacillus	Large, gram-positive bacillus.	Aerobes or facultative anaerobes; form endo-spores; B. anthracis causes anthrax in cattle and humans, B. cereus may cause food poison-ing.
Clostridium	Gram-positive bacillus.	Strict anaerobes; form endospores; all known species are pathogenic, causing tetanus, gas gangrene, botulism, and colitis.

Enterococcus	Gram-positive coccus; forms microscopic pairs in culture (resembling Streptococcus pneumoniae).	Anaerobic aerotolerant bacteria, abundant in the human gut, may cause urinary tract and other infections in the nosocomial environment.
Lactobacillus	Gram-positive bacillus.	Facultative anaerobes; ferment sugars into lactic acid; part of the vaginal microbiota; used as probiotics.
Leuconostoc	Gram-positive coccus; may form microscopic chains in culture	Fermenter, used in food industry to produce sauerkraut and kefir.
Mycoplasma	The smallest bacteria; appear pleomorphic under electron microscope.	Have no cell wall; classified as low G+C Gram-positive bacteria because of their genome; M. pneumoniae causes "walking" pneumonia.
Staphylococcus	Gram-positive coccus; forms microscopic clusters in culture that resemble bunches of grapes.	Tolerate high salt concentration; facultative anaerobes; produce catalase; S. aureus can also produce coagulase and toxins responsible for local (skin) and generalized infections.
Streptococcus	Gram-positive coccus; forms chains or pairs in culture.	Diverse genus; classified into groups based on sharing certain antigens; some species cause hemolysis and may produce toxins responsible for human local (throat) and generalized disease.
Ureaplasma	Similar to Mycoplasma.	Part of the human vaginal and lower urinary tract microbiota; may cause inflammation, sometimes leading to internal scarring and infertility.

METHICILLIN-RESISTANT STAPHYLOCOCCUS AUREUS

Methicillin-resistant Staphylococcus aureus (MRSA) refers to a group of gram-positive bacteria that are genetically distinct from other strains of Staphylococcus aureus. MRSA is responsible for several difficult-to-treat infections in humans. MRSA is any strain of S. aureus that has developed, through horizontal gene transfer and natural selection, multiple drug resistance to beta-lactam antibiotics. β-lactam antibiotics are a broad spectrum group which includes some penams – penicillin derivatives such as methicillin and oxacillin, and cephems such as the cephalosporins. Strains unable to resist these antibiotics are classified as methicillin-susceptible S. aureus, or MSSA.

MRSA is common in hospitals, prisons, and nursing homes, where people with open wounds, invasive devices such as catheters, and weakened immune systems are at greater risk of hospital-acquired infection. MRSA began as a hospital-acquired infection, but has become community-acquired as well as livestock-acquired. The terms HA-MRSA (healthcare-associated or hospital-acquired MRSA), CA-MRSA (community-associated MRSA) and LA-MRSA (livestock-associated) reflect this.

Signs and Symptoms

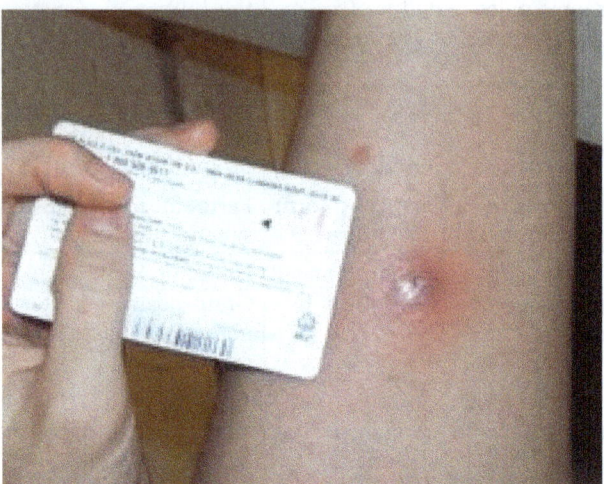

Although usually carried without symptoms, MRSA often presents as
small red pustular skin infections.

In humans, S. aureus is part of the normal microbiota present in the upper respiratory
tract, and on skin and in the gut mucosa. S. aureus, along with similar species that can
colonize and act symbiotically but can cause disease if they begin to take over the tissues they have colonized or invade other tissues, have been called "pathobionts".

After 72 hours, MRSA can take hold in human tissues and eventually become resistant
to treatment. The initial presentation of MRSA is small red bumps that resemble pimples, spider bites, or boils; they may be accompanied by fever and, occasionally, rashes.
Within a few days, the bumps become larger and more painful; they eventually open
into deep, pus-filled boils. About 75 percent of CA-MRSA infections are localized to
skin and soft tissue and usually can be treated effectively.

Risk Factors

A select few of the populations at risk:

- People with indwelling implants, prostheses, drains, and catheters.

- People who are frequently in crowded places, especially with shared equipment
  and skin-to-skin contact.

- People with weak immune systems (HIV/AIDS, lupus, or cancer sufferers;
  transplant recipients; severe asthmatics; etc.)

- Diabetics.

- Intravenous drug users.

- Users of quinolone antibiotics.

- Elderly people.

- School children sharing sports and other equipment.

- College students living in dormitories.

- People staying or working in a health care facility for an extended period of time.

- People who spend time in coastal waters where MRSA is present, such as some beaches in Florida and the west coast of the United States.

- People who spend time in confined spaces with other people, including occupants of homeless shelters, prison inmates, and military recruits in basic training.

- Veterinarians, livestock handlers, and pet owners.

- People that ingest unpasteurized milk.

- People who are immunocompromised and also colonized.

- People with Chronic obstructive pulmonary disease.

- People who had thoracic surgery.

As many as 22 percent of people infected with MRSA do not have any discernable risk factors.

Hospitalized People

People who are hospitalized, including the elderly, are often immunocom promised and susceptible to infection of all kinds, including MRSA; when the infection is by MRSA this is called healthcare-associated or hospital-acquired methicillin-resistant S. aureus (HA-MRSA). Generally, those infected by MRSA will stay infected for just under 10 days, if treated by a doctor, although effects may vary from person to person.

Surgical as well as nonsurgical wounds can be infected with HA-MRSA. Surgical site infections (SSI) occur on the skin surface but can spread to internal organs and blood to cause sepsis. Transmission occurs between healthcare providers and patients. This is because some providers may inconsistently neglect to perform hand-washing between examinations.

People in nursing homes are at risk for all the reasons above, further complicated by the generally weaker immune systems of the elderly or other residents in need of such care.

Prison Inmates and Military Recruits

Prisons, and military barracks, can be crowded and confined, and poor hygiene practices may proliferate, thus putting inhabitants at increased risk of contracting MRSA.

Cases of MRSA in such populations were first reported in the United States, and then in Canada. The earliest reports were made by the Center for Disease Control (CDC) in US state prisons. In the news media, hundreds of reports of MRSA outbreaks in prisons appeared between 2000 and 2008. For example, in February 2008, the Tulsa County jail in Oklahoma started treating an average of 12 *S. aureus* cases per month.

Animals

Antibiotic use in livestock increases the risk that MRSA will develop among the livestock; strains MRSA ST 398 and CC398 are transmissible to humans. Generally, animals are asymptomatic.

Domestic pets are susceptible to MRSA infection from their owners; MRSA infected pets can also transmit MRSA to humans.

Athletes

Locker rooms, gyms, and related athletic facilities offer potential sites for MRSA contamination and infection. Athletes have been identified as a high risk group. A study linked MRSA to the abrasions caused by artificial turf. Three studies by the Texas State Department of Health found the infection rate among football players was 16 times the national average. In October 2006, a high-school football player was temporarily paralyzed from MRSA-infected turf burns. His infection returned in January 2007 and required three surgeries to remove infected tissue, as well as three weeks of hospital stay.

In 2013, Lawrence Tynes, Carl Nicks, and Johnthan Banks of the Tampa Bay Buccaneers were diagnosed with MRSA. Tynes and Nicks apparently did not contract the infection from each other, but it is unknown if Banks contracted it from either individual. In 2015, Los Angeles Dodgers' infielder Justin Turner was infected while the team visited the New York Mets. In October 2015, New York Giants tight end Daniel Fells was hospitalized with a serious MRSA infection.

Children

MRSA is becoming a critical problem in children; studies found 4.6% of patients in U.S. health-care facilities, (presumably) including hospital nurseries, were infected or colonized with MRSA. Children (and adults, as well) who come in contact with day-care centers, playgrounds, locker rooms, camps, dormitories, classrooms and other school settings, and gyms and workout facilities are at higher risk of getting MRSA. Parents should be especially cautious of children who participate in activities where sports equipment is shared, such as football helmets and uniforms.

Mechanism

Antimicrobial resistance is genetically based; resistance is mediated by the acquisition of extrachromosomal genetic elements containing resistance genes. Examples include plasmids, transposable genetic elements, and genomic islands, which are transferred between bacteria through horizontal gene transfer. A defining characteristic of MRSA is its ability to thrive in the presence of penicillin-like antibiotics, which normally prevent bacterial growth by inhibiting synthesis of cell wall material. This is due to a resistance gene, mecA, which stops β-lactam antibiotics from inactivating the enzymes (transpeptidases) critical for cell wall synthesis.

SCCmec

Staphylococcal cassette chromosome mec (SCCmec) is a genomic island of unknown origin containing the antibiotic resistance gene mecA. SCCmec contains additional genes beyond mecA, including the cytolysin gene psm-mec, which may suppress virulence in HA-acquired MRSA strains. In addition this locus encodes strain dependent gene regulatory RNA called psm-mecRNA. SCCmec also contains ccrA and ccrB; both genes encode recombinases that mediate the site-specific integration and excision of the SCCmec element from the S. aureus chromosome. Currently, six unique SCCmec types ranging in size from 21–67 kb have been identified; they are designated types I-VI and are distinguished by variation in mec and ccr gene complexes. Owing to the size of the SCCmec element and the constraints of horizontal gene transfer, a minimum of five clones are thought to be responsible for the spread of MRSA infections, with clonal complex (CC) 8 most prevalent. SCCmec is thought to have originated in the closely related S. sciuri species and transferred horizontally to S. aureus.

Different SCCmec genotypes confer different microbiological characteristics, such as different antimicrobial resistance rates. Different genotypes are also associated with different types of infections. Types I-III SCCmec are large elements that typically contain additional resistance genes and are characteristically isolated from HA-MRSA strains. Conversely, CA-MRSA is associated with types IV and V, which are smaller and lack resistance genes other than mecA.

These distinctions were thoroughly investigated by Collins et al. in 2001 and can be explained by the fitness differences associated with carriage of a large or small SCC-mec plasmid. Carriage of large plasmids, such as SCCmecI-III, is costly to the bacteria, resulting in compensatory decrease in virulence expression. MRSA is able to thrive in hospital settings with increased antibiotic resistance but decreased virulence- HA-MRSA targets immunocompromised, hospitalized hosts, thus a decrease in virulence is not maladaptive. In contrast, CA-MRSA tends to carry lower fitness cost SCCmec elements to offset the increased virulence and toxicity expression required to infect healthy hosts.

mecA

mecA is a biomarker gene responsible for resistance to methicillin and other β-lactam antibiotics. After acquisition of mecA, the gene must be integrated and localized in the

S. aureus chromosome. mecA encodes penicillin-binding protein 2a (PBP2a), which differs from other penicillin-binding proteins as its active site does not bind methicillin or other β-lactam antibiotics. As such, PBP2a can continue to catalyze the transpeptidation reaction required for peptidoglycan cross-linking, enabling cell wall synthesis in the presence of antibiotics. As a consequence of the inability of PBP2a to interact with β-lactam moieties, acquisition of mecA confers resistance to all β-lactam antibiotics in addition to methicillin.

mecA is under the control of two regulatory genes, mecI and mecR1. MecI is usually bound to the mecA promoter and functions as a repressor. In the presence of a β-lactam antibiotic, MecR1 initiates a signal transduction cascade that leads to transcriptional activation of mecA. This is achieved by MecR1-mediated cleavage of MecI, which alleviates MecI repression. mecA is further controlled by two co-repressors, BlaI and BlaR1. blaI and blaR1 are homologous to mecI and mecR1, respectively, and normally function as regulators of blaZ, which is responsible for penicillin resistance. The DNA sequences bound by MecI and BlaI are identical; therefore, BlaI can also bind the mecA operator to repress transcription of mecA.

Arginine Catabolic Mobile Element

The arginine catabolic mobile element (ACME) is a virulence factor present in many MRSA strains but not prevalent in MSSA. SpeG-positive ACME compensates for the polyamine hypersensitivity of S. aureus and facilitates stable skin colonization, wound infection, and person-to-person transmission.

Strains

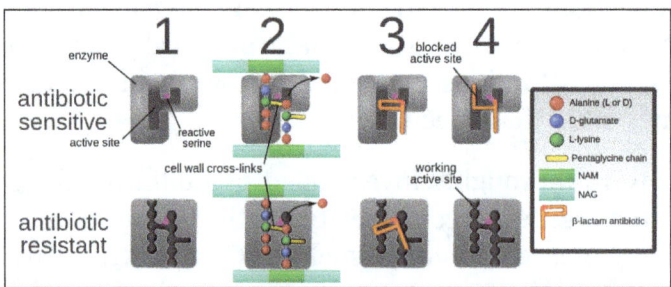

Depicting antibiotic resistance through alteration of the antibiotic's target site, modeled after MRSA's resistance to penicillin.

Beta-lactam antibiotics permanently inactivate PBP enzymes, which are essential for bacterial life, by permanently binding to their active sites. Some forms of MRSA, however, express a PBP that will not allow the antibiotic into their active site.

Acquisition of SCCmec in methicillin-sensitive S. aureus (MSSA) gives rise to a number of genetically different MRSA lineages. These genetic variations within different MRSA strains possibly explain the variability in virulence and associated MRSA infections. The first MRSA strain, ST250 MRSA-1 originated from SCCmec and ST250-MSSA integration. Historically, major MRSA clones: ST2470-MRSA-I, ST239-MRSA-III, ST5-MRSA-II, and ST5-MRSA-IV were responsible for causing hospital-acquired MRSA (HA-MRSA) infections. ST239-MRSA-III, known as the Brazilian clone, was highly transmissible compared to others and distributed in Argentina, Czech Republic, and Portugal.

In the UK, the most common strains of MRSA are EMRSA15 and EMRSA16. EMRSA16 has been found to be identical to the ST36:USA200 strain, which circulates in the United States, and to carry the SCCmec type II, enterotoxin A and toxic shock syndrome toxin 1 genes. Under the new international typing system, this strain is now called MRSA252. EMRSA 15 is also found to be one of the common MRSA strains in Asia. Other common strains include ST5:USA100 and EMRSA 1. These strains are genetic characteristics of HA-MRSA.

Community-acquired MRSA (CA-MRSA) strains emerged in late 1990 to 2000, infecting healthy people who had not been in contact with health care facilities. Researchers suggest that CA-MRSA did not evolve from the HA-MRSA. This is further proven by molecular typing of CA-MRSA strains and genome comparison between CA-MRSA and HA-MRSA, which indicate that novel MRSA strains integrated SCCmec into MSSA separately on its own. By mid 2000, CA-MRSA was introduced into the health care systems and distinguishing CA-MRSA from HA-MRSA became a difficult process. Community-acquired MRSA is more easily treated and more virulent than hospital-acquired MRSA (HA-MRSA). The genetic mechanism for the enhanced virulence in CA-MRSA remains an active area of research. Especially the Panton–Valentine leukocidin (PVL) genes are of interest because they are a unique feature of CA-MRSA.

In the United States, most cases of CA-MRSA are caused by a CC8 strain designated ST8:USA300, which carries SCCmec type IV, Panton–Valentine leukocidin, PSM-alpha and enterotoxins Q and K, and ST1:USA400. The ST8:USA300 strain results in skin infections, necrotizing fasciitis and toxic shock syndrome, whereas the ST1:USA400 strain results in necrotizing pneumonia and pulmonary sepsis. Other community-acquired strains of MRSA are ST8:USA500 and ST59:USA1000. In many nations of the world, MRSA strains with different predominant genetic background types have come to predominate among CA-MRSA strains; USA300 easily tops the list in the U.S. and is becoming more common in Canada after its first appearance there in 2004. For example, in Australia ST93 strains are common, while in continental Europe ST80 strains, which carry SCCmec type IV, predominate. In Taiwan, ST59 strains, some of which are resistant to many non-beta-lactam antibiotics, have arisen as common causes of skin and soft tissue infections in the community. In a remote region of Alaska, unlike most of the continental U.S., USA300 was found rarely in a study of MRSA strains from outbreaks in 1996 and 2000 as well as in surveillance from 2004–06.

A MRSA strain, CC398, is found in intensively reared production animals (primarily pigs, but also cattle and poultry), where it can be transmitted to humans as LA-MRSA (livestock-associated MRSA).

Diagnosis

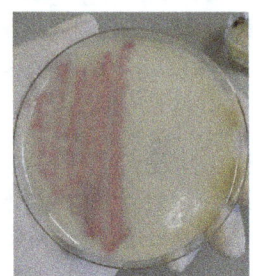

A selective and differential chromogenic medium for the qualitative direct detection of MRSA.

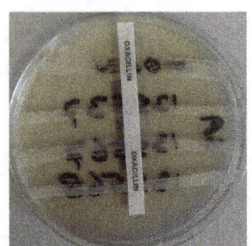

The MRSA resistance to oxacillin being tested, the top s. aureus isolate is control and sensitive to oxacillin, the other three isolates are MRSA positive.

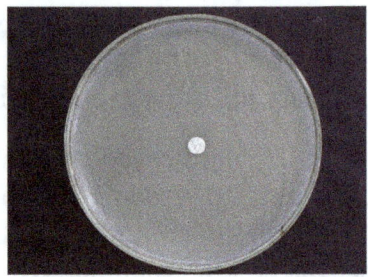

Mueller Hinton agar showing MRSA resistant to oxacillin disk.

Diagnostic microbiology laboratories and reference laboratories are key for identifying outbreaks of MRSA. Normally, the bacterium must be cultured from blood, urine, sputum, or other body-fluid samples, and in sufficient quantities to perform confirmatory tests early-on. Still, because no quick and easy method exists to diagnose MRSA, initial treatment of the infection is often based upon 'strong suspicion' and techniques by the treating physician; these include quantitative PCR procedures, which are employed in clinical laboratories for quickly detecting and identifying MRSA strains.

Another common laboratory test is a rapid latex agglutination test that detects the PB-P2a protein. PBP2a is a variant penicillin-binding protein that imparts the ability of *S. aureus* to be resistant to oxacillin.

Microbiology

All S. aureus (also abbreviated SA at times), methicillin-resistant S. aureus (MRSA) is a gram-positive, spherical (coccus) bacterium that is about 1 micron in diameter. It does not form spores and it is non-motile. It forms grape-like clusters or chains. Unlike Methicillin-susceptible S. aureus (MSSA), MRSA is slower growing on a variety of media and has been found to exist in mixed colonies of MSSA. The mecA gene, which confers the resistance to a number of antibiotics is present in MRSA and not in MSSA. In some instances, the mecA gene is present in MSSA but is not expressed. Polymerase chain reaction (PCR) testing is the most precise method in identifying MRSA strains. Specialized culture media have been developed to better differentiate between MSSA and MRSA and in some cases, it will identify specific strains that are resistant to different antibiotics.

Other strains of S. aureus have emerged that are resistant to oxacillin, clindamycin, teicoplanin, and erythromycin. These resistant strains may or may not possess the mecA gene. S. aureus has also developed resistance to vancomycin (VRSA). One strain is only partially susceptible to vancomycin and is called vancomycin-intermediate S. aureus (VISA). GISA is a strain of resistant S. aureus and stands for glycopeptide-intermediate

S. aureus and is less suspectible to vancomycin and teicoplanin. Resistance to antibiotics in S. aureus can be quantified. This done by determining the amount of the antibiotic in micrograms/milliliter must be used to inhibit growth. If S. aureus is inhibited at a concentration of vancomycin of less than or equal to 4 micrograms/milliliter, it is said to be susceptible. If a concentration of greater than 32 micrograms/milliliter is necessary to inhibit growth, it is said to be resistant.

Prevention

Screening

In health care settings, isolating those with MRSA from those without the infection is one method to prevent transmission. Rapid culture and sensitivity testing and molecular testing identifies carriers and reduces infection rates.

MRSA can be identified by swabbing the nostrils and isolating the bacteria found inside the nostrils. Combined with extra sanitary measures for those in contact with infected people, swab screening people admitted to hospitals has been found to be effective in minimizing the spread of MRSA in hospitals in the United States, Denmark, Finland, and the Netherlands.

Hand Washing

The CDC offers suggestions for preventing the contraction and spread MRSA infection which are applicable to those in community settings, including incarcerated populations, childcare center employees, and athletes. To prevent the spread of MRSA the recommendations are to wash hands using soap and water or an alcohol-based sanitizer. Additional recommendations are to keep wounds clean and covered, avoid contact with other people's wounds, avoid sharing personal items such as razors or towels, shower after exercising at athletic facilities, and shower before using swimming pools or whirlpools.

Isolation

Excluding medical facilities, current US guidance does not require workers with MRSA infections to be routinely excluded from the general workplace. The National Institutes of Health recommends that those with wound drainage that cannot be covered and contained with a clean, dry bandage and those who cannot maintain good hygiene practices be reassigned. Workers with active infections are excluded from activities where skin-to-skin contact is likely to occur. To prevent the spread of staph or MRSA in the workplace, employers make available adequate facilities that encourage good hygiene. In addition, surface and equipment sanitizing conforms to the Environmental Protection Agency (EPA)-registered disinfectants.

Health Departments recommend that preventing the spread of MRSA in the home can be to: launder materials that have come into contact with infected person separately and with a dilute bleach solution; reduce the bacterial load in your nose and on your skin; clean those things in the house that people regularly touch like sinks, tubs, kitchen counters, cell phones, light switches, doorknobs, phones, toilets, and computer keyboards.

In hospital settings, once one to three cultures come back negative, contact isolation can be stopped.

Restricting Antibiotic Use

Glycopeptides, cephalosporins, and, in particular, quinolones are associated with an increased risk of colonisation of MRSA. Reducing use of antibiotic classes that promote MRSA colonisation, especially fluoroquinolones, is recommended in current guidelines.

Public Health Considerations

Mathematical models describe one way in which a loss of infection control can occur after measures for screening and isolation seem to be effective for years, as happened in the UK. In the "search and destroy" strategy that was employed by all UK hospitals until the mid-1990s, all hospitalized people with MRSA were immediately isolated, and all staff were screened for MRSA and were prevented from working until they had completed a course of eradication therapy that was proven to work. Loss of control occurs because colonised people are discharged back into the community and then readmitted; when the number of colonised people in the community reaches a certain threshold, the "search and destroy" strategy is overwhelmed. One of the few countries not to have been overwhelmed by MRSA is the Netherlands: An important part of the success of the Dutch strategy may have been to attempt eradication of carriage upon discharge from hospital.

Decolonization

As of 2013 there had been no randomized clinical trials conducted to understand how to treat non-surgical wounds that had been colonized, but not infected, with MRSA, and insufficient studies had been conducted to understand how to treat surgical wounds that had been colonized with MRSA. As of 2013 it was not known whether strategies to eradicate MRSA colonization of people in nursing homes reduced infection rates.

Care should be taken when trying to drain boils, as disruption of surrounding tissue can lead to larger infections, or even infection of the blood stream (often with fatal consequences).

Mupirocin 2% ointment can be effective at reducing the size of lesions. A secondary covering of clothing is preferred. As shown in an animal study with diabetic mice, the topical application of a mixture of sugar (70%) and 3% povidone-iodine paste is an effective agent for the treatment of diabetic ulcers with MRSA infection.

Community Settings

It may be difficult for people to maintain the necessary cleanliness if they do not have access to facilities such as public toilets with handwashing facilities. In the United Kingdom, the Workplace (Health, Safety and Welfare) Regulations 1992 requires businesses to provide toilets for their employees, along with washing facilities including soap or other suitable means of cleaning. Guidance on how many toilets to provide and what sort of washing facilities should be provided alongside them is given in the Workplace (Health, Safety and Welfare) Approved Code of Practice and Guidance L24, available from Health and Safety Executive Books. But there is no legal obligation on local authorities in the United Kingdom to provide public toilets, and although in 2008 the House of Commons Communities and Local Government Committee called for a duty on local authorities to develop a public toilet strategy this was rejected by the Government.

Agriculture

Some advocate regulations on the use of antibiotics in animal food to prevent the emergence of drug resistant strains of MRSA. MRSA is established in animals and birds.

Treatment

Antibiotics

Treatment is urgent and delays can be fatal. The location and history related to the infection determines the treatment. The route of administration of an antibiotic varies. Antibiotics effective against MRSA can be given by IV, oral, or a combination of both and depends on the specific circumstances and patient characteristics. The use of concurrent treatment with vancomycin other beta-lactam agents may have a synergistic effect.

Both CA-MRSA and HA-MRSA are resistant to traditional anti-staphylococcal beta-lactam antibiotics, such as cephalexin. CA-MRSA has a greater spectrum of antimicrobial susceptibility to sulfa drugs (like co-trimoxazole (trimethoprim/sulfamethoxazole), tetracyclines (like doxycycline and minocycline) and clindamycin (for osteomyelitis). MRSA can be eradicated with a regimen of linezolid, though treatment protocols vary and serum levels of antibiotics vary widely person to person and may affect outcomes. The effective treatment of MRSA with linezolid has been successful in 87% of people. Linezolid is more effective in soft tissue infections than vancomycin. This is compared to eradication of infection in those with MRSA treated with vancomycin. Treatment with vancomycin is successful in approximately 49% of people. Linezolid belongs to the newer oxazolidinone class of antibiotics which has been shown to be effective against both CA-MRSA and HA-MRSA. The Infectious Disease Society of America recommends vancomycin, linezolid, or clindamycin (if susceptible) for treating those with MRSA pneumonia. Ceftaroline, a fifth-generation cephalosporin, is the first beta-lactam antibiotic approved in the US to treat MRSA infections in skin and soft tissue or community acquired pneumonia.

Vancomycin and teicoplanin are glycopeptide antibiotics used to treat MRSA infections. Teicoplanin is a structural congener of vancomycin that has a similar activity spectrum but a longer half-life. Because the oral absorption of vancomycin and teicoplanin is very low, these agents can be administered intravenously to control systemic infections. Treatment of MRSA infection with vancomycin can be complicated, due to its inconvenient route of administration. Moreover, the efficacy of vancomycin against MRSA is inferior to that of anti-staphylococcal beta-lactam antibiotics against methicillin-susceptible S. aureus (MSSA).

Several newly discovered strains of MRSA show antibiotic resistance even to vancomycin and teicoplanin. These new strains of the MRSA bacterium have been dubbed vancomycin intermediate-resistant S. aureus (VISA). Linezolid, quinupristin/dalfopristin, daptomycin, ceftaroline, and tigecycline are used to treat more severe infections that do not respond to glycopeptides such as vancomycin. Current guidelines recommend daptomycin for VISA bloodstream infections and endocarditis.

This left vancomycin as the only effective agent available at the time. However, strains with intermediate (4–8 µg/ml) levels of resistance, termed glycopeptide-intermediate S. aureus (GISA) or vancomycin-intermediate S. aureus (VISA), began appearing in the late 1990s. The first identified case was in Japan in 1996, and strains have since been found in hospitals in England, France and the US. The first documented strain with complete (>16 µg/ml) resistance to vancomycin, termed vancomycin-resistant S. aureus (VRSA) appeared in the United States in 2002. However, in 2011, a variant of vancomycin has been tested that binds to the lactate variation and also binds well to the original target, thus reinstating potent antimicrobial activity.

Oxazolidinones such as linezolid, became available in the 1990s, and are comparable to vancomycin in effectiveness against MRSA. Linezolid resistance in S. aureus was reported in 2001, but infection rates have been at consistently low levels and in the United Kingdom and Ireland, no resistance was found in staphylococci collected from bacteremia cases between 2001 and 2006.

Skin and Soft-tissue Infections

In skin abscesses, the primary treatment recommended is removal of dead tissue, incision, and drainage. More data is needed to determine the effectiveness of specific antibiotics therapy in SSIs. Examples of soft tissue infections from MRSA include: ulcers, impetigo, abscesses, and surgical site infections.

In surgical wounds, there is weak evidence (high risk of bias) that Linezolid may be better than Vancomycin to eradicate MRSA surgical site infections.

MRSA colonization is also found in non surgical wounds such as traumatic wounds, burns, and chronic ulcers (i.e.: diabetic ulcer, pressure ulcer, arterial insufficiency ulcer, venous ulcer). There is no conclusive evidence about the best antibiotic regimen to treat MRSA colonization.

Children

In skin infections and in secondary infection sites topical mupirocin is used successfully. For bacteremia and endocarditis, vancomycin or daptomycin is considered. For children with MRSA infected bone or joints, treatment is individualized and long-term. Neonates can develop Neonatal pustulosis as a result of topical infection with MRSA. Clindamycin is not approved for the treatment of MRSA infection it is still used in children for soft tissue infections.

Endocarditis and Bacteremia

Evaluation for the replacement of a prosthetic valve is considered. Appropriate antibiotic therapy may be administered for up to six weeks. Four to six weeks of antibiotic treatment is often recommended, and is dependent upon the extent of MRSA infection.

Respiratory Infections

CA-MRSA in hospitalized patients pneumonia treatment begins before culture results. After the susceptibility to antibiotics is performed, the infection may be treated with vancomycin or linezolid for up to 21 days. If the pneumonia is complicated by the accumulation of pus in the pleural cavity surrounding the lungs, drainage may be done along with antibiotic therapy. People with cystic fibrosis may develop respiratory complications related to MRSA infection. The incidence of MRSA in those with cystic fibrosis increased during 2000 to 2015 by five times. Most of these infections were HA-MRSA. MRSA accounts for 26% of lung infections in those with cystic fibrosis.

Bone and Joint Infections

Cleaning the wound of dead tissue and draining abscesses is the first action to treat the MRSA infection. Administration of antibiotics is not standardized and is adapted by a case-by-case basis. Antibiotic therapy can last up to 1 to 3 months and sometimes even longer.

Infected Implants

MRSA infection can occur associated with implants and joint replacements. Recommendations on treatment are based upon the length of time the implant has been in place. In cases of a recent placement of a surgical implant or artificial joint, the device may be retained while antibiotic therapy continues. If the placement of the device has occurred over 3 weeks ago, the device may be removed. Antibiotic therapy is used in each instance sometimes long-term.

Central Nervous System

MRSA can infect the central nervous system and form brain abscess, subdural empyema, and spinal epidural abscess. Excision and drainage can be done along with antibiotic treatment. Septic thrombosis of cavernous or dural venous sinus can sometimes be a complication.

Other Infections

Treatment is not standardized for other instances of MRSA infection in a wide range of tissues. Treatment varies for MRSA infections related to: subperiosteal abscesses, necrotizing pneumonia, cellulitis, pyomyositis, necrotizing fasciitis, mediastinitis, myocardial, perinephric, hepatic, and splenic abscesses, septic thrombophlebitis, and severe ocular infections, including endophthalmitis. Pets can be reservoirs and pass on MRSA to people. In some cases, the infection can be symptomatic and the pet can suffer a MRSA infection. Health departments recommend that the pet be taken to the veterinarian if MRSA infections keep occurring in the people who have contact with the pet.

Epidemiology

Worldwide, an estimated 2 billion people carry some form of S. aureus; of these, up to 53 million (2.7% of carriers) are thought to carry MRSA.

HA-MRSA

In a US cohort study of 1300 healthy children, 2.4% carried MRSA in their nose. Bacterial sepsis occurs with most (75%) of cases of invasive MRSA infection. In 2009, there were an estimated 463,017 hospitalization due to MRSA or a rate of 11.74 per 1,000 hospitalizations. Many of these infections are less serious, but the Centers for Disease Control and Prevention (CDC) estimates that there are 80,461 invasive MRSA infections and 11,285 deaths due to MRSA annually. In 2003, the cost for a hospitalization due to a MRSA was $92,363. A hospital stay for MSSA was $52,791 (USD).

Infection after surgery is relatively uncommon, but occurs as much as 33% in specific types of surgeries. Infections of surgical sites range from 1% to 33%. MRSA sepsis that occurs within 30 days has a 15-38% mortality rate. MRSA sepsis that occurs within one year following a surgical infection has a mortality rate of around 55%. There may be increased mortality associated with cardiac surgery. There is a rate of 12.9% in those infected with MRSA while only a 3% infected with other organisms. SSIs infected with MRSA had longer hospital stays than those who did not.

Globally, MRSA infection rates are dynamic and vary year to year. According to The 2006 SENTRY Antimicrobial Surveillance Program report, the incidence of MRSA blood stream infections was 35.9 percent in North America. MRSA blood infections in Latin America was 29%. European incidence was 22.8%. The rate of all MRSA infections in Europe ranged from 50% percent in Portugal down to 0.8 per cent in Sweden. Overall MRSA infection rates varied in Latin America: Colombia and Venezuela combined had 3%, Mexico had 50%, Chile 38%, Brazil 29%, and Argentina 28%.

The Centers for Disease Control and Prevention (CDC) estimated that about 1.7 million nosocomial infections occurred in the United States in 2002, with 99,000 associated deaths. The estimated incidence is 4.5 nosocomial infections per 100 admissions, with direct costs (at 2004 prices) ranging from $10,500 per case (for bloodstream, urinary tract, or respiratory infections in immunocompetent people) to $111,000 (£57,000, €85,000) per case for antibiotic-resistant infections in the bloodstream in people with transplants. With these numbers, conservative estimates of the total direct costs of nosocomial infections are above $17 billion. The reduction of such infections forms an important component of efforts to improve healthcare safety. MRSA alone was associated with 8% of nosocomial infections reported to the CDC National Healthcare Safety Network from January 2006 to October 2007.

The British National Audit Office estimated that the incidence of nosocomial infections in Europe ranges from 4% to 10% of all hospital admissions. As of early 2005, the number of deaths in the United Kingdom attributed to MRSA has been estimated by various sources to lie in the area of 3,000 per year.

In the United States, 95 million carry S. aureus in their noses; of these, 2.5 million (2.6% of carriers) carry MRSA. A population review conducted in three U.S. communities showed the annual incidence of CA-MRSA during 2001–2002 to be 18–25.7/100,000; most CA-MRSA isolates were associated with clinically relevant infections, and 23% of people required hospitalization.

CA-MRSA

There are concerns that the presence of MRSA in the environment may allow resistance to be transferred to other bacteria through phages (viruses that infect bacteria). The source of MRSA could come from hospital waste, farm sewage, and other waste water.

STAPHYLOCOCCUS EPIDERMIDIS

Staphylococcus epidermidis is a Gram-positive bacterium, and one of over 40 species belonging to the genus Staphylococcus. It is part of the normal human flora, typically the skin flora, and less commonly the mucosal flora. It is a facultative anaerobic bacteria. Although S. epidermidis is not usually pathogenic, patients with compromised immune systems are at risk of developing infection. These infections are generally hospital-acquired. S. epidermidis is a particular concern for people with catheters or other surgical implants because it is known to form biofilms that grow on these devices. Being part of the normal skin flora, S. epidermidis is a frequent contaminant of specimens sent to the diagnostic laboratory.

Friedrich Julius Rosenbach distinguished S. epidermidis from S. aureus in 1884, initially naming S. epidermidis as S. albus. He chose aureus and albus since the bacteria formed yellow and white colonies, respectively.

Cellular Morphology and Biochemistry

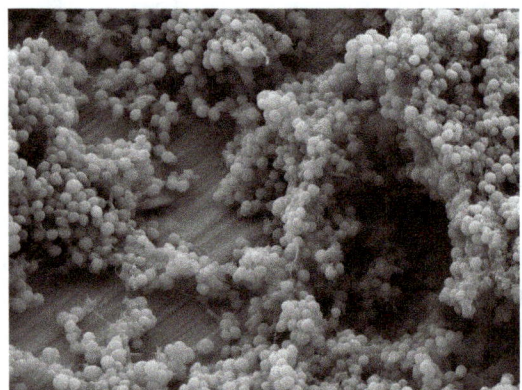

Staphylococcus epidermidis biofilm on titanium substrate.

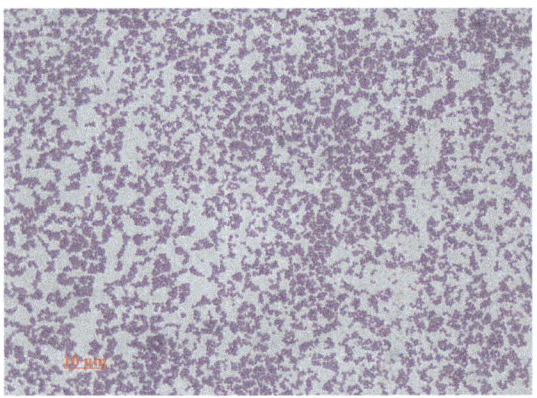

Staphylococcus epidermidis, 1000 magnification under bright field microscopy.

S. epidermidis is a very hardy microorganism, consisting of nonmotile, Gram-positive cocci, arranged in grape-like clusters. It forms white, raised, cohesive colonies about 1–2 mm in diameter after overnight incubation, and is not hemolytic on blood agar. It is a catalase-positive, coagulase-negative, facultative anaerobe that can grow by aerobic respiration or by fermentation. Some strains may not ferment.

Biochemical tests indicate this microorganism also carries out a weakly positive reaction to the nitrate reductase test. It is positive for urease production, is oxidase negative, and can use glucose, sucrose, and lactose to form acid products. In the presence of lactose, it will also produce gas. S. epidermidis does not possess the gelatinase enzyme, so it cannot hydrolyze gelatin. It is sensitive to novobiocin, providing an important test to distinguish it from Staphylococcus saprophyticus, which is coagulase-negative, as well, but novobiocin-resistant.

Similar to those of S. aureus, the cell walls of S. epidermidis have a transferrin-binding protein that helps the organism obtain iron from transferrin. The tetramers of a surface exposed protein, glyceraldehyde-3-phosphate dehydrogenase, are believed to bind to transferrin and remove its iron. Subsequent steps include iron being transferred to surface lipoproteins, then to transport proteins which carry the iron into the cell.

Virulence and Antibiotic Resistance

The ability to form biofilms on plastic devices is a major virulence factor for S. epidermidis. One probable cause is surface proteins that bind blood and extracellular matrix proteins. It produces an extracellular material known as polysaccharide intercellular adhesin (PIA), which is made up of sulfated polysaccharides. It allows other bacteria to bind to the already existing biofilm, creating a multilayer biofilm. Such biofilms decrease the metabolic activity of bacteria within them. This decreased metabolism, in combination with impaired diffusion of antibiotics, makes it difficult for antibiotics to effectively clear this type of infection. S. epidermidis strains are often resistant to antibiotics, including rifamycin, fluoroquinolones, gentamicin, tetracycline, clindamycin, and sulfonamides. Methicillin resistance is particularly widespread, with 75-90% of hospital isolates resistance to methicilin. Resistant organisms are most commonly found in the intestine, but organisms living freely on the skin can also become resistant due to routine exposure to antibiotics secreted in sweat.

Disease

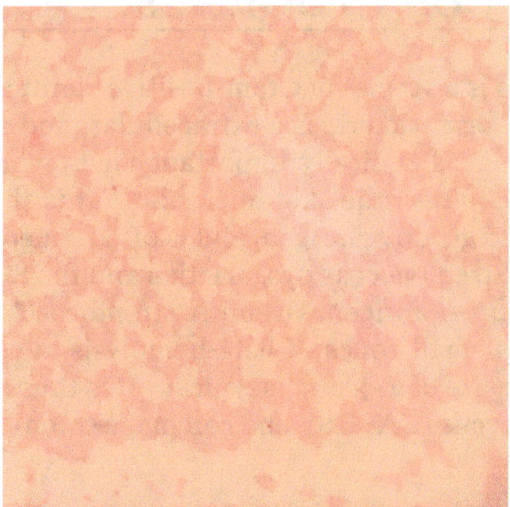
Staphylococcus epidermidis stained by safranin.

S. epidermidis causes biofilms to grow on plastic devices placed within the body. This occurs most commonly on intravenous catheters and on medical prostheses. Infection can also occur in dialysis patients or anyone with an implanted plastic device that may have been contaminated. It also causes endocarditis, most often in patients with defective heart valves. In some other cases, sepsis can occur in hospital patients.

Antibiotics are largely ineffective in clearing biofilms. The most common treatment for these infections is to remove or replace the infected implant, though in all cases, prevention is ideal. The drug of choice is often vancomycin, to which rifampin or an aminoglycoside can be added. Hand washing has been shown to reduce the spread of infection.

Preliminary research also indicates S. epidermidis is universally found inside affected acne vulgaris pores, where Cutibacterium acnes is normally the sole resident.

Identification

The normal practice of detecting S. epidermidis is by using appearance of colonies on selective media, bacterial morphology by light microscopy, catalase and slide coagulase testing. On the Baird-Parker agar with egg yolk supplement, colonies appear small and black. Increasingly, techniques such as quantitative PCR are being employed for the rapid detection and identification of Staphylococcus strains. Normally, sensitivity to desferrioxamine can also be used to distinguish it from most other staphylococci, except in the case of Staphylococcus hominis, which is also sensitive. In this case, the production of acid from trehalose by S. hominis can be used to tell the two species apart.

STREPTOCOCCUS PYOGENES

Streptococcus pyogenes is a species of Gram-positive, aerotolerant bacterium in the genus Streptococcus. These bacteria are extracellular, and made up of non-motile and non-sporing cocci. It is clinically important for humans. It is an infrequent, but usually pathogenic, part of the skin microbiota. It is the predominant species harboring the Lancefield group A antigen, and is often called group A streptococcus (GAS). However, both Streptococcus dysgalactiae and the Streptococcus anginosus group can possess group A antigen. Group A streptococci when grown on blood agar typically produces small zones of beta-hemolysis, a complete destruction of red blood cells. (A zone size of 2–3 mm is typical.) It is thus also called group A (beta-hemolytic) streptococcus (GABHS), and can make colonies greater than 5 mm in size.

Like other cocci, streptococci are round bacteria because streptococcal cells tend to link in chains of round cells and a number of infections caused by the bacterium produce pus. The main criterion for differentiation between Staphylococcus spp. and Streptococcus spp. is the catalase test. Staphylococci are catalase positive whereas streptococci are catalase-negative. S. pyogenes can be cultured on fresh blood agar plates. Under ideal conditions, it has an incubation period of 1 to 3 days.

An estimated 700 million GAS infections occur worldwide each year. While the overall mortality rate for these infections is 0.1%, over 650,000 of the cases are severe and invasive, and have a mortality rate of 25%. Early recognition and treatment are critical; diagnostic failure can result in sepsis and death.

Epidemiology

S. pyogenes typically colonises the throat, genital mucosa, rectum, and skin. Of healthy individuals, 1% to 5% have throat, vaginal, or rectal carriage. In healthy children, such carriage rate varies from 2% to 17%. There are four methods for the transmission of this bacterium: inhalation of respiratory droplets, skin contact, contact with objects, surface, or dust that is contaminated with bacteria or, less commonly, transmission through food. Such bacteria can cause a variety of diseases such as streptococcal pharyngitis, rheumatic fever, rheumatic heart disease, and scarlet fever. Although pharyngitis is mostly viral in origin, about 15 to 30% of all pharyngitis cases in children are caused by GAS; meanwhile, 5 to 20% of pharyngitis in adults are streptococcal. The number of pharyngitis cases is higher in children when compared with adults due to exposures in schools, nurseries, and as a consequence of lower host immunity. Such cases Streptococcal pharyngitis occurs more frequently from December to April (later winter to early spring) in seasonal countries, possibly due to changing climate, behavioural changes or predisposing viral infection. Disease cases are the lowest during autumn.

MT1 (metabolic type 1) clone is frequently associated with invasive Streptococcus pyogenes infections among developed countries. The incidence and mortality of S. pyognes was high during the pre-penicillin era, but had already started to fall prior to the widespread availability of penicillin. Therefore, environmental factors do play a role in the S. pyogenes infection. Incidence of S. pyogenes is 2 to 4 per 100,000 population in developed countries and 12 to 83 per 100,000 population in developing countries. S. pyogenes infection is more frequently found in men than women, with highest rates in the elderly, followed by infants. In people with risk factors such as heart disease, diabetes, malignancy, blunt trauma, surgical incision, virus respiratory infection, including influenza, S. pyogenes infection happens in 17 to 25% of all cases. GAS secondary infection usually happens within one week of the diagnosis of influenza infection. In 14 to 16% of childhood S. pyogenes infections, there is a prior chickenpox infection. Such S. pyogenes infection in children usually manifests as severe soft tissue infection with onset 4 to 12 days from the chickenpox diagnosis. There is also 40 to 60 times increase in risk of S. pyogenes infection within the first two weeks of chickenpox infection in children. However, 20 to 30% of S. pyogenes infection does occur in adults with no identifiable risk factors. The incidence is higher in children (50 to 80% of S. pyogenes infection) with no known risk factors. The rates of scarlet fever in UK was usually 4 in 100,000 population, however, in 2014, the rates had risen to 49 per 100,000 population. Rheumatic fever and rheumatic heart disease (RHD) usually occurs at 2 to 3 weeks after the throat infection, which is more common among the impoverished people in developing countries. From 1967 to 1996, the global mean incidence of rheumatic fever and RHD was 19 per 100,000 with the highest incidence at 51 per 100,000.

Maternal S. pyogenes infection usually happens in late pregnancy; at more than 30 weeks of gestation to four weeks post partum, which accounts for 2 to 4% of all the S. pyogenes infections. This represents 20 to 100 times increase in risk for S. pyogenes infections. Clinical manifestations are: pneumonia, septic arthritis, necrotizing fasciitis, and genital tract sepsis. According to a study done by Queen Charlotte's hospital in London during the 1930s, the vagina was not the common source of such infection. On the contrary, maternal throat infection and close contacts with carriers were the more common sites for maternal S. pyogenes infection.

Bacteriology

Serotyping

In 1928, Rebecca Lancefield published a method for serotyping S. pyogenes based on its cell-wall polysaccharide, a virulence factor displayed on its surface. Later, in 1946, Lancefield described the serologic classification of S. pyogenes isolates based on their surface T-antigen. Four of the 20 T-antigens have been revealed to be pili, which are used by bacteria to attach to host cells. As of 2016, a total of 120 M proteins are identified. These M proteins are encoded by 234 types emm gene with greater than 1,200 alleles.

Lysogeny

All strains of S. pyogenes are polylysogenized, in that they carry one or more bacteriophage on their genomes. Some of the 'phages may be defective, but in some cases active 'phage may compensate for defects in others. In general, the genome of S. pyogenes strains isolated during disease are >90% identical, they differ by the 'phage they carry.

Virulence Factors

S. pyogenes has several virulence factors that enable it to attach to host tissues, evade the immune response, and spread by penetrating host tissue layers. A carbohydrate-based bacterial capsule composed of hyaluronic acid surrounds the bacterium, protecting it from phagocytosis by neutrophils. In addition, the capsule and several factors embedded in the cell wall, including M protein, lipoteichoic acid, and protein F (SfbI) facilitate attachment to various host cells. M protein also inhibits opsonization by the alternative complement pathway by binding to host complement regulators. The M protein found on some serotypes is also able to prevent opsonization by binding to fibrinogen. However, the M protein is also the weakest point in this pathogen's defense, as antibodies produced by the immune system against M protein target the bacteria for engulfment by phagocytes. M proteins are unique to each strain, and identification can be used clinically to confirm the strain causing an infection.

- Streptolysin O: An exotoxin, one of the bases of the organism's beta-hemolytic property, streptolysin O causes an immune response and detection of antibodies to it; antistreptolysin O (ASO) can be clinically used to confirm a recent infection. It is damaged by oxygen.

- Streptolysin S: A cardiotoxic exotoxin, another beta-hemolytic component, not immunogenic and O_2 stable: A potent cell poison affecting many types of cell including neutrophils, platelets, and subcellular organelles.

- Streptococcal Pyrogenic Exotoxin A (Spea)/Streptococcal Pyrogenic Exotoxin C (SpeC): Superantigens secreted by many strains of S. pyogenes: This pyrogenic exotoxin is responsible for the rash of scarlet fever and many of the symptoms of streptococcal toxic shock syndrome, also known as toxic shock like syndrome(TSLS).

- Streptokinase: Enzymatically activates plasminogen, a proteolytic enzyme, into plasmin, which in turn digests fibrin and other proteins.

- Hyaluronidase: Hyaluronidase is widely assumed to facilitate the spread of the bacteria through tissues by breaking down hyaluronic acid, an important component of connective tissue. However, very few isolates of S. pyogenes are capable of secreting active hyaluronidase due to mutations in the gene that encode the enzyme. Moreover, the few isolates capable of secreting hyaluronidase do not appear to need it to spread through tissues or to cause skin lesions. Thus, the true role of hyaluronidase in pathogenesis, if any, remains unknown.

- Streptodornase: Most strains of S. pyogenes secrete up to four different DNases, which are sometimes called streptodornase. The DNases protect the bacteria from being trapped in neutrophil extracellular traps (NETs) by digesting the NETs' web of DNA, to which are bound neutrophil serine proteases that can kill the bacteria.

- C5a Peptidase: C5a peptidase cleaves a potent neutrophil chemotaxin called C5a, which is produced by the complement system. C5a peptidase is necessary to minimize the influx of neutrophils early in infection as the bacteria are attempting to colonize the host's tissue. C5a peptidase, although required to degrade the neutrophil chemotaxin C5a in the early stages of infection, is not required for S. pyogenes to prevent the influx of neutrophils as the bacteria spread through the fascia.

- Streptococcal Chemokine Protease: The affected tissue of patients with severe cases of necrotizing fasciitis are devoid of neutrophils. The serine protease ScpC, which is released by S. pyogenes, is responsible for preventing the migration of neutrophils to the spreading infection. ScpC degrades the chemokine IL-8, which would otherwise attract neutrophils to the site of infection.

Genome

The genome of different strains were sequenced (genome size is 1.8–1.9 Mbp) encoding about 1700-1900 proteins (1700 in strain NZ131, 1865 in strain MGAS5005).

Disease

S. pyogenes is the cause of many human diseases, ranging from mild superficial skin infections to life-threatening systemic diseases. Infections typically begin in the throat or skin. The most striking sign is a strawberry-like rash. Examples of mild S. pyogenes infections include pharyngitis (strep throat) and localized skin infection (impetigo). Erysipelas and cellulitis are characterized by multiplication and lateral spread of S. pyogenes in deep layers of the skin. S. pyogenes invasion and multiplication in the fascia can lead to necrotizing fasciitis, a life-threatening condition requiring surgery. The bacterium is found in neonatal infections.

Infections due to certain strains of S. pyogenes can be associated with the release of bacterial toxins. Throat infections associated with release of certain toxins lead to scarlet fever. Other toxigenic S. pyogenes infections may lead to streptococcal toxic shock syndrome, which can be life-threatening.

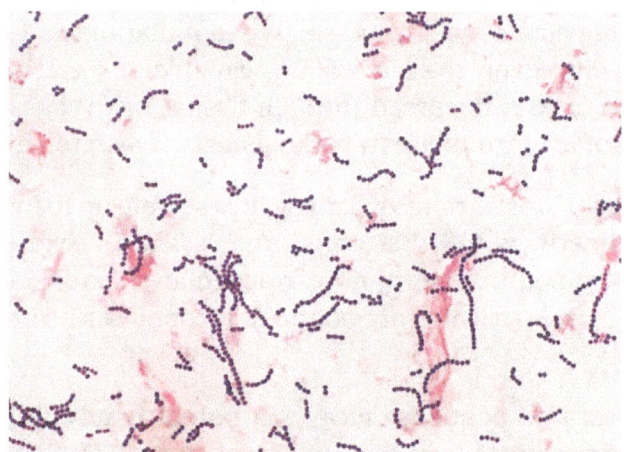

S. pyogenes can also cause disease in the form of post-infectious "non-pyogenic" (not associated with local bacterial multiplication and pus formation) syndromes. These autoimmune-mediated complications follow a small percentage of infections and include rheumatic fever and acute post-infectious glomerulonephritis. Both conditions appear several weeks following the initial streptococcal infection. Rheumatic fever is characterized by inflammation of the joints and/or heart following an episode of streptococcal pharyngitis. Acute glomerulonephritis, inflammation of the renal glomerulus, can follow streptococcal pharyngitis or skin infection.

This bacterium remains acutely sensitive to penicillin. Failure of treatment with penicillin is generally attributed to other local commensal organisms producing β-lactamase, or failure to achieve adequate tissue levels in the pharynx. Certain strains have developed resistance to macrolides, tetracyclines, and clindamycin.

Applications

Bionanotechnology

Many S. pyogenes proteins have unique properties, which have been harnessed in recent years to produce a highly specific "superglue" and a route to enhance the effectiveness of antibody therapy.

Genome Editing

The CRISPR system from this organism that is used to recognize and destroy DNA from invading viruses, stopping the infection, was appropriated in 2012 for use as a genome-editing tool that could potentially alter any piece of DNA and later RNA.

BACILLUS

Bacillus is a genus of Gram-positive, rod-shaped bacteria, a member of the phylum Firmicutes, with 266 named species. The term is also used to describe the shape (rod) of certain bacteria; and the plural Bacilli is the name of the class of bacteria to which this genus belongs. Bacillus species can be either obligate aerobes: oxygen dependent; or facultative anaerobes: having the ability to be anaerobic in the absence of oxygen. Cultured Bacillus species test positive for the enzyme catalase if oxygen has been used or is present.

Bacillus can reduce themselves to oval endospores and can remain in this dormant state for years. The endospore of one species from Morocco is reported to have survived being heated to 420 °C. Endospore formation is usually triggered by a lack of nutrients: the bacterium divides within its cell wall, and one side then engulfs the other. They are not true spores (i.e., not an offspring). Endospore formation originally defined the genus, but not all such species are closely related, and many species have been moved to other genera of the Firmicutes.

Ubiquitous in nature, Bacillus includes both free-living (nonparasitic) species, and two parasitic pathogenic species. These two Bacillus species are medically significant: B. anthracis causes anthrax; and B. cereus causes food poisoning.

Many species of Bacillus can produce copious amounts of enzymes which are used in various industries, such as the production of alpha amylase used in starch hydrolysis, and the protease subtilisin used in detergents. B. subtilis is a valuable model for bacterial research. Some Bacillus species can synthesize and secrete lipopeptides, in particular surfactins and mycosubtilins.

Structure

Cell Wall

The cell wall of Bacillus is a structure on the outside of the cell that forms the second barrier between the bacterium and the environment, and at the same time maintains the rod shape and withstands the pressure generated by the cell's turgor. The cell wall is composed of teichoic and teichuronic acids. B. subtilis is the first bacterium for which the role of an actin-like cytoskeleton in cell shape determination and peptidoglycan synthesis was identified, and for which the entire set of peptidoglycan-synthesizing enzymes was localised. The role of the cytoskeleton in shape generation and maintenance is important.

Naming

The genus Bacillus was named in 1835 by Christian Gottfried Ehrenberg, to contain rod-shaped (bacillus) bacteria. He had seven years earlier named the genus Bacterium. Bacillus was later amended by Ferdinand Cohn to further describe them as spore-forming, Gram-positive, aerobic or facultatively anaerobic bacteria. Like other genera associated with the early history of microbiology, such as Pseudomonas and Vibrio, the 266 species of Bacillus are ubiquitous. The genus has a very large ribosomal 16S diversity.

Isolation and Identification

An easy way to isolate Bacillus species is by placing nonsterile soil in a test tube with water, shaking, placing in melted mannitol salt agar, and incubating at room temperature for at least a day. Cultured colonies are usually large, spreading, and irregularly shaped.

Under the microscope, the Bacillus cells appear as rods, and a substantial portion of the cells usually contain oval endospores at one end, making them bulge.

Phylogeny

Three proposals have been presented as representing the phylogeny of the genus Bacillus. The first proposal, presented in 2003, is a Bacillus-specific study, with the most diversity covered using 16S and the ITS regions. It divides the genus into 10 groups. This includes the nested genera Paenibacillus, Brevibacillus, Geobacillus, Marinibacillus and Virgibacillus.

The second proposal, presented in 2008, constructed a 16S (and 23S if available) tree of all validated species. The genus Bacillus contains a very large number of nested taxa and majorly in both 16S and 23S. It is paraphyletic to the Lactobacillales (Lactobacillus, Streptococcus, Staphylococcus, Listeria, etc.), due to Bacillus coahuilensis and others.

A third proposal, presented in 2010, was a gene concatenation study, and found results similar to the 2008 proposal, but with a much more limited number of species in terms of groups.(This scheme used Listeria as an outgroup, so in light of the ARB tree, it may be "inside-out").

One clade, formed by B. anthracis, B. cereus, B. mycoides, B. pseudomycoides, B. thuringiensis, and B. weihenstephanensis under the 2011 classification standards, should be a single species (within 97% 16S identity), but due to medical reasons, they are considered separate species (an issue also present for four species of Shigella and Escherichia coli).

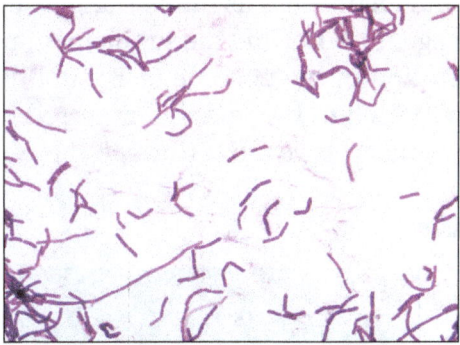

Phylogeny of the genus Bacillus.

Ecological Significance

Bacillus species are ubiquitous in nature, e.g. in soil. They can occur in extreme environments such as high pH (B. alcalophilus), high temperature (B. thermophilus), and high salt concentrations (B. halodurans). B. thuringiensis produces a toxin that can kill insects and thus has been used as insecticide. B. siamensis has antimicrobial compounds that inhibit plant pathogens, such as the fungi Rhizoctonia solani and Botrytis cinerea, and they promote plant growth by volatile emissions. Some species of Bacillus are naturally competent for DNA uptake by transformation.

Clinical Significance

- Two *Bacillus* species are medically significant: *B. anthracis*, which causes anthrax; and *B. cereus*, which causes food poisoning, with symptoms similar to that caused by *Staphylococcus*.

- *B. thuringiensis* is an important insect pathogen, and is sometimes used to control insect pests.

- *B. subtilis* is an important model organism. It is also a notable food spoiler, causing ropiness in bread and related food.

- Some environmental and commercial strains of *B. coagulans* may play a role in food spoilage of highly acidic, tomato-based products.

Industrial Significance

Many Bacillus species are able to secrete large quantities of enzymes. Bacillus amylo-liquefaciens is the source of a natural antibiotic protein barnase (a ribonuclease), alpha amylase used in starch hydrolysis, the protease subtilisin used with detergents, and the BamH1 restriction enzyme used in DNA research. A portion of the Bacillus thuringiensis genome was incorporated into corn (and cotton) crops. The resulting GMOs are resistant to some insect pests. Bacillus species continue to be dominant bacterial workhorses in microbial fermentations. Bacillus subtilis (natto) is the key microbial participant in the ongoing production of the soya-based traditional natto fermentation Bacillus strains have also been developed and engineered as industrial producers of nucleotides, the vitamin riboflavin, the flavor agent ribose, and the supplement poly-gamma-glutamic acid. With the recent characterization of the genome of B. subtilis 168 and of some related strains, Bacillus species are poised to become the preferred hosts for the production of many new and improved products as we move through the genomic and proteomic era.

Use as Model Organism

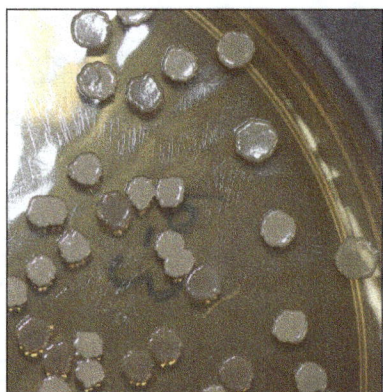

Colonies of the model species Bacillus subtilis on an agar plate.

Bacillus subtilis is one of the best understood prokaryotes, in terms of molecular and cellular biology. Its superb genetic amenability and relatively large size have provided the powerful tools required to investigate a bacterium from all possible aspects. Recent improvements in fluorescent microscopy techniques have provided novel insight into the dynamic structure of a single cell organism. Research on B. subtilis has been at the forefront of bacterial molecular biology and cytology, and the organism is a model for differentiation, gene/protein regulation, and cell cycle events in bacteria.

SPIRAL BACTERIA

Spiral bacteria, bacteria of spiral (helical) shape, form the third major morphological category of prokaryotes along with the rod-shaped bacilli and round cocci. Spiral bacteria can be subclassified by the number of twists per cell, cell thickness, cell flexibility, and motility. The two types of spiral cells are spirillum and spirochete, with spirillum being rigid with external flagella, and spirochetes being flexible with internal flagella.

Spirillum

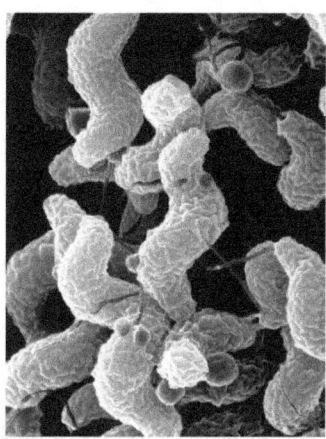

Campylobacter jejuni is a common spirrillum bacterium.

A spirillum (plural spirilla) is a rigid spiral bacterium that is Gram-negative and frequently has external amphitrichous or lophotrichous flagella. Examples include:

- Members of the genus Spirillum.

- Campylobacter species, such as Campylobacter jejuni, a foodborne pathogen that causes campylobacteriosis.

- Helicobacter species, such as Helicobacter pylori, a cause of peptic ulcers.

Spirochete

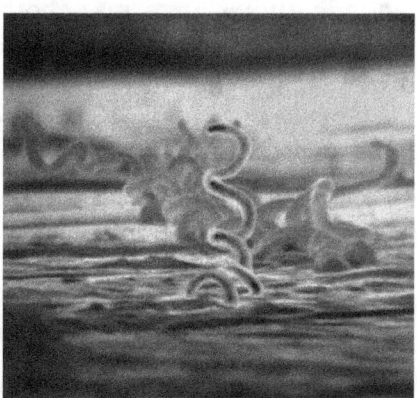

Thin spirochete Treponema pallidum bacteria, the causative agent of syphilis magnified 400 times.

A spirochete (plural spirochetes) is a very thin, elongate, flexible, spiral bacteria that is motile via internal periplasmic flagella inside the outer membrane. Owing to their morphological properties, spirochetes are difficult to Gram-stain but may be visualized using dark field microscopy or Warthin–Starry stain. Examples include:

- Members of the phylum Spirochaetes.

- Leptospira species, which cause leptospirosis.

- Borrelia species, such as Borrelia burgdorferi, a tick-borne bacterium that causes Lyme disease.

- Treponema species, such as Treponema pallidum, subspecies of which causes treponematoses, including syphilis.

CUTIBACTERIUM ACNES

Cutibacterium acnes (formerly Propionibacterium acnes) is the relatively slow-growing, typically aerotolerant anaerobic, Gram-positive bacterium (rod) linked to the skin condition of acne; it can also cause chronic blepharitis and endophthalmitis, the latter particularly following intraocular surgery. Its genome has been sequenced and a study has shown several genes can generate enzymes for degrading skin and proteins that may be immunogenic (activating the immune system).

The species is largely commensal and part of the skin flora present on most healthy adult humans' skin. It is usually just barely detectable on the skin of healthy preadolescents. It lives, among other things, primarily on fatty acids in sebum secreted by sebaceous glands in the follicles. It may also be found throughout the gastrointestinal tract.

Originally identified as Bacillus acnes, it was later named Propionibacterium acnes for its ability to generate propionic acid. In 2016, P. acnes was taxonomically reclassified as a result of biochemical and genomic studies. In terms of both phylogenetic tree structure and DNA G + C content, the cutaneous species was distinguishable from other species that had been previously categorized as P. acnes. As part of restructuring, the novel genus Cutibacterium was created for the cutaneous species, including those formerly identified as Propionibacterium acnes, Propionibacterium avidum, and Propionibacterium granulosum. Characterization of phylotypes of C. acnes is an active field of research.

Role in Disease

C. acnes bacteria predominantly live deep within follicles and pores, although they are also found on the surface of healthy skin. In these follicles, C. acnes bacteria use sebum, cellular debris and metabolic byproducts from the surrounding skin tissue as their primary sources of energy and nutrients. Elevated production of sebum by hyperactive sebaceous glands (sebaceous hyperplasia) or blockage of the follicle can cause C. acnes bacteria to grow and multiply.

C. acnes bacteria secrete many proteins, including several digestive enzymes. These enzymes are involved in the digestion of sebum and the acquisition of other nutrients. They can also destabilize the layers of cells that form the walls of the follicle. The cellular damage, metabolic byproducts and bacterial debris produced by the rapid growth of C. acnes in follicles can trigger inflammation. This inflammation can lead to the symptoms associated with some common skin disorders, such as folliculitis and acne vulgaris.

The damage caused by C. acnes and the associated inflammation make the affected tissue more susceptible to colonization by opportunistic bacteria, such as Staphylococcus aureus. Preliminary research shows healthy pores are only colonized by C. acnes, while unhealthy ones universally include the nonpore-resident Staphylococcus epidermidis, amongst other bacterial contaminants. Whether this is a root causality, just opportunistic and a side effect, or a more complex pathological duality between C. acnes and this particular Staphylococcus species is not known.

C. acnes has also been found in corneal ulcers, and is a common cause of chronic endophthalmitis following cataract surgery. Rarely, it infects heart valves leading to endocarditis, and infections of joints (septic arthritis) have been reported. Furthermore, Cutibacterium species have been found in ventriculostomy insertion sites, and areas subcutaneous to suture sites in patients who have undergone craniotomy. It is a common contaminant in blood and cerebrospinal fluid cultures.

C. acnes has been found in herniated discs. The propionic acid which it secretes creates micro-fractures of the surrounding bone. These micro-fractures are sensitive and it has been found that antibiotics have been helpful in resolving this type of low back pain.

C. acnes can be found in bronchoalveolar lavage of approximately 70% of patients with sarcoidosis and is associated with disease activity, but it can be also found in 23% of controls. The subspecies of C. acnes that cause these infections of otherwise sterile tissues (prior to medical procedures), however, are the same subspecies found on the skin of individuals who do not have acne-prone skin, so are likely local contaminants. Moderate to severe acne vulgaris appears to be more often associated with virulent strains.

C. acnes is an opportunistic pathogen, causing a range of postoperative and device-related infections e.g., surgery, post-neurosurgical infection, joint prostheses, shunts and prosthetic heart valves. C. acnes may play a role in other conditions, including SAPHO (synovitis, acne, pustulosis, hyperostosis, osteitis) syndrome, sarcoidosis and sciatica. It is also suspected a main bacterial source of neuroinflammation in Alzheimer's disease brains.

Antimicrobial Susceptibility

C. acnes bacteria are susceptible to a wide range of antimicrobial molecules, from both pharmaceutical and natural sources. Antibiotics are commonly used to treat infections caused by C. acnes. Acne vulgaris is the disease most commonly associated with C. acnes infection. The antibiotics most frequently used to treat acne vulgaris are erythromycin, clindamycin, doxycycline, and minocycline. Several other families of antibiotics are also active against C. acnes bacteria, including quinolones, cephalosporins, pleuromutilins, penicillins, and sulfonamides.

The emergence of antibiotic-resistant C. acnes bacteria represents a growing problem worldwide. The problem is especially pronounced in North America and Europe. The antibiotic families that C. acnes are most likely to acquire resistance to are the macrolides (e.g., erythromycin and azithromycin), lincosamides (e.g., clindamycin) and tetracyclines (e.g., doxycycline and minocycline).

However, C. acnes bacteria are susceptible to many types of antimicrobial chemicals found in over-the-counter antibacterial products, including benzoyl peroxide, triclosan, chloroxylenol, and chlorhexidine gluconate.

Several naturally occurring molecules and compounds are toxic to C. acnes bacteria. Some essential oils such as rosemary, tea tree oil, clove oil, and citrus oils contain antibacterial chemicals.

The elements silver, sulfur, and copper have also been demonstrated to be toxic towards many bacteria, including C. acnes. Natural honey has also been shown to have some antibacterial properties that may be active against C. acnes.

Photosensitivity

C. acnes glows orange when exposed to blacklight, possibly due to the presence of endogenous porphyrins. It is also killed by ultraviolet light. C. acnes is especially sensitive to light in the 405–420 nanometer (near the ultraviolet) range due to an endogenic porphyrin–coporphyrin III. A total irradiance of 320 Joules/cm^2 inactivates this species in vitro. Its photosensitivity can be enhanced by pretreatment with aminolevulinic acid, which boosts production of this chemical, although this causes significant side effects in humans, and in practice was not significantly better than the light treatment alone.

Other Habitats

C. acnes has been found to be an endophyte of plants. Notably, grapevine appears to host an endophytic population of C. acnes that is closely related to the human-associated strains. The two lines diverged roughly 7,000 years ago, at about the same time when grapevine agriculture may have been established. This C. acnes subtype was dubbed Zappae in honour of the eccentric composer Frank Zappa, to highlight its unexpected and unconventional habitat.

Corynebacterium

Corynebacterium is a genus of bacteria that are Gram-positive and aerobic. They are bacilli (rod-shaped), and in some phases of life they are, more particularly, club-shaped, which inspired the genus name (coryneform means "club-shaped").

They are widely distributed in nature in the microbiota of animals (including the human microbiota) and are mostly innocuous, most commonly existing in commensal relationships with their hosts. Some are useful in industrial settings such as C. glutamicum. Others can cause human disease, including most notably diphtheria, which is caused by C. diphtheriae. As with various species of a microbiota (including their cousins in the genera Arcanobacterium and Trueperella), they usually are not pathogenic but can occasionally opportunistically capitalize on atypical access to tissues (via wounds) or weakened host defenses.

Taxonomy

The genus Corynebacterium was created by Lehmann and Neumann in 1896 as a taxonomic group to contain the bacterial rods responsible for causing diphtheria. The genus was defined based on morphological characteristics. Based on studies of 16S-rRNA, they have been grouped into the subdivision of gram-positive eubacteria with high G:C content, with close phylogenetic relationship to Arthrobacter, Mycobacterium, Nocardia, and Streptomyces.

The term "diphtheroids" is used to represent corynebacteria that are non-pathogenic; for example, C. diphtheriae would be excluded.

Genomics

Comparative analysis of corynebacterial genomes has led to the identification of several conserved signature indels which are unique to the genus. Two examples of these conserved signature indels are a two-amino-acid insertion in a conserved region of the enzyme phosphoribose diphosphate: decaprenyl-phosphate phosphoribosyltransferase and a three-amino-acid insertion in acetate kinase, both of which are found only in Corynebacterium species. Both of these indels serve as molecular markers for species of the genus Corynebacterium. Additionally, 16 conserved signature proteins, which are uniquely found in Corynebacterium species, have been identified. Three of the conserved signature proteins have homologs found in the genus Dietzia, which is believed to be the closest related genus to Corynebacterium. In phylogenetic trees based on concatenated protein sequences or 16S rRNA, the genus Corynebacterium forms a distinct clade, within which is a distinct subclade, cluster I. The cluster is made up of the species C. diptheriae, C. pseudotuberculosis, C. ulcerans, C. aurimucosum, C. glutamicum, and C. efficiens. This cluster is distinguished by several conserved signature indels, such as a two-amino-acid insertion in LepA and a seven- or eight-amino-acid insertions in RpoC. Also, 21 conserved signature proteins are found only in members of cluster I. Another cluster has been proposed, consisting of C. jeikeium and C. urealyticum, which is supported by the presence of 19 distinct conserved signature proteins which are unique to these two species. Corynebateria have a high G+C content ranging from 46-74 mol%.

Characteristics

The principal features of the genus Corynebacterium were described by Collins and Cummins in 1986. They are gram-positive, catalase-positive, non-spore-forming, non-motile, rod-shaped bacteria that are straight or slightly curved. Metachromatic granules are usually present representing stored phosphate regions. Their size falls between 2 and 6 μms in length and 0.5 μm in diameter. The bacteria group together in a characteristic way, which has been described as the form of a "V", "palisades", or "Chinese characters". They may also appear elliptical. They are aerobic or facultatively anaerobic, chemoorganotrophs. They are pleomorphic through their lifecycles, they occur in various lengths, and they frequently have thickenings at either end, depending on the surrounding conditions.

Cell Wall

The cell wall is distinctive, with a predominance of mesodiaminopimelic acid in the murein wall and many repetitions of arabinogalactan, as well as corynemycolic acid (a mycolic acid with 22 to 26 carbon atoms), bound by disaccharide bonds called L-Rhap-(1 → 4)--D-GlcNAc-phosphate. These form a complex commonly seen in Corynebacterium species: the mycolyl-AG–peptidoglican (mAGP).

Culture

Corynebacteria grow slowly, even on enriched media. In terms of nutritional requirements, all need biotin to grow. Some strains also need thiamine and PABA. Some of the Corynebacterium species with sequenced genomes have between 2.5 and 3.0 million base pairs. The bacteria grow in Loeffler's medium, blood agar, and trypticase soy agar (TSA). They form small, grayish colonies with a granular appearance, mostly translucent, but with opaque centers, convex, with continuous borders. The color tends to be yellowish-white in Loeffler's medium. In TSA, they can form grey colonies with black centers and dentated borders that look similar to flowers (C. gravis), or continuous borders (C. mitis), or a mix between the two forms (C. intermedium).

Habitat

Corynebacterium species occur commonly in nature in the soil, water, plants, and food products. The nondiphtheiroid Corynebacterium species can even be found in the mucosa and normal skin flora of humans and animals. Unusual habitats, such as the preen gland of birds have been recently reported for Corynebacterium uropygiale. Some species are known for their pathogenic effects in humans and other animals. Perhaps the most notable one is C. diphtheriae, which acquires the capacity to produce diphtheria toxin only after interacting with a bacteriophage. Other pathogenic species in humans include: C. amycolatum, C. striatum, C. jeikeium, C. urealyticum, and C. xerosis; all of these are important as pathogens in immunosuppressed patients. Pathogenic species in other animals include C. bovis and C. renale. This genus has been found to be part of the human salivary microbiome.

Role in Disease

The most notable human infection is diphtheria, caused by C. diphtheriae. It is an acute and contagious infection characterized by pseudomembranes of dead epithelial cells, white blood cells, red blood cells, and fibrin that form around the tonsils and back of the throat. In developed countries, it is an uncommon illness that tends to occur in unvaccinated individuals, especially school-aged children, elderly, neutropenic or immunocompromised patients, and those with prosthetic devices such as prosthetic heart valves, shunts, or catheters. It is more common in developing countries It can occasionally infect wounds, the vulva, the conjunctiva, and the middle ear. It can be spread within a hospital. The virulent and toxigenic strains are lysogenic, and produce an exotoxin formed by two polypeptide chains, which is itself produced when a bacterium is transformed by a gene from the β prophage.

Several species cause disease in animals, most notably C. pseudotuberculosis, which causes the disease caseous lymphadenitis, and some are also pathogenic in humans. Some attack healthy hosts, while others tend to attack the immunocompromised. Effects of infection include granulomatous lymphadenopathy, pneumonitis, pharyngitis, skin infections, and endocarditis. Corynebacterial endocarditis is seen most frequently in patients with intravascular devices. C. tenuis is believed to cause trichomycosis palmellina and trichomycosis axillaris. C. striatum may cause axillary odor. C. minutissimum causes erythrasma.

Industrial Uses

Nonpathogenic species of Corynebacterium are used for very important industrial applications, such as the production of amino acids, nucleotides, and other nutritional factors; bioconversion of steroids; degradation of hydrocarbons; cheese aging; and production of enzymes. Some species produce metabolites similar to antibiotics: bacteriocins of the corynecin-linocin type, antitumor agents, etc. One of the most studied species is C. glutamicum, whose name refers to its capacity to produce glutamic acid in aerobic conditions. This is used in the food industry as monosodium glutamate in the production of soy sauce and yogurt.

Species of Corynebacterium have been used in the mass production of various amino acids including glutamic acid, a food additive that is made at a rate of 1.5 million tons/year. The metabolic pathways of Corynebacterium have been further manipulated to produce lysine and threonine.

L-Lysine production is specific to C. glutamicum in which core metabolic enzymes are manipulated through genetic engineering to drive metabolic flux towards the production of NADPH from the pentose phosphate pathway, and L-4-aspartyl phosphate, the commitment step to the synthesis of L-lysine, lysC, dapA, dapC, and dapF. These enzymes are up-regulated in industry through genetic engineering to ensure adequate amounts of lysine precursors are produced to increase metabolic flux. Unwanted side reactions such as threonine and asparagine production can occur if a buildup of intermediates occurs, so scientists have developed mutant strains of C. glutamicum through PCR engineering and chemical knockouts to ensure production of side-reaction enzymes are limited. Many genetic manipulations conducted in industry are by traditional cross-over methods or inhibition of transcriptional activators.

Expression of functionally active human epidermal growth factor has been brought about in C. glutamicum, thus demonstrating a potential for industrial-scale production of human proteins. Expressed proteins can be targeted for secretion through either the general secretory pathway or the twin-arginine translocation pathway.

Unlike gram-negative bacteria, the gram-positive Corynebacterium species lack lipopolysaccharides that function as antigenic endotoxins in humans.

Species

Nonlipophilic

The nonlipophilic bacteria may be classified as fermentative and nonfermentative:

- Fermentative corynebacteria:
    - Corynebacterium diphtheriae group
    - Corynebacterium xerosis and Corynebacterium striatum
    - Corynebacterium minutissimum
    - Corynebacterium amycolatum
    - Corynebacterium glucuronolyticum
    - Corynebacterium argentoratense
    - Corynebacterium matruchotii
    - Corynebacterium glutamicum
    - Corynebacterium sp.
- Nonfermentative corynebacteria:
    - Corynebacterium afermentans subsp. afermentans
    - Corynebacterium auris
    - Corynebacterium pseudodiphtheriticum
    - Corynebacterium propinquum

- Lipophilic:
  - Corynebacterium uropygiale
  - Corynebacterium jeikeium
  - Corynebacterium urealyticum
  - Corynebacterium afermentans subsp. lipophilum
  - Corynebacterium accolens
  - Corynebacterium macginleyi
  - CDC coryneform groups F-1 and G
  - Corynebacterium bovis
- Novel Corynebacteria that do not contain mycolic acids:
  - Corynebacterium kroppenstedtii.

GRAM POSITIVE VS GRAM NEGATIVE BACTERIA

Gram-positive bacteria are those species with peptidoglycan outer layers, are easier to kill - their thick peptidoglycan layer absorbs antibiotics and cleaning products easily. In contrast, their many-membraned cousins resist this intrusion with their multi-layered structure. Therefore, infection prevention techniques must ensure that they can breach the thick peptidoglycan layer of the Gram-positive bacteria but also get through the many layers of the Gram-negative bacteria.

However thin their peptidoglycan layer, Gram-negative bacteria are protected from certain physical assaults because they do not absorb foreign materials that surround it (including Gram's purple dye). Imagine a spacecraft with a series of airlocks. Any intruder would have to make their way through these airlocks before entering the ship. Such is the case with gram-negative bacteria. Their additional membrane allows them to control what reaches the inner airlock, enabling them to sequester or even remove threats in that space between the membranes (periplasmic space) before it reaches the cell itself.

As a result, Gram-negative bacteria are not destroyed by certain detergents which easily kill Gram-positive bacteria. While thick, the Gram-positive bacteria's membrane absorbs foreign materials (Gram's dye), even those that prove toxic to its insides. This makes them easier to destroy with certain detergents. As a result, only certain cleansers are approved for use to eliminate bacteria - because it must kill both Gram-positive and Gram-negative bacteria.

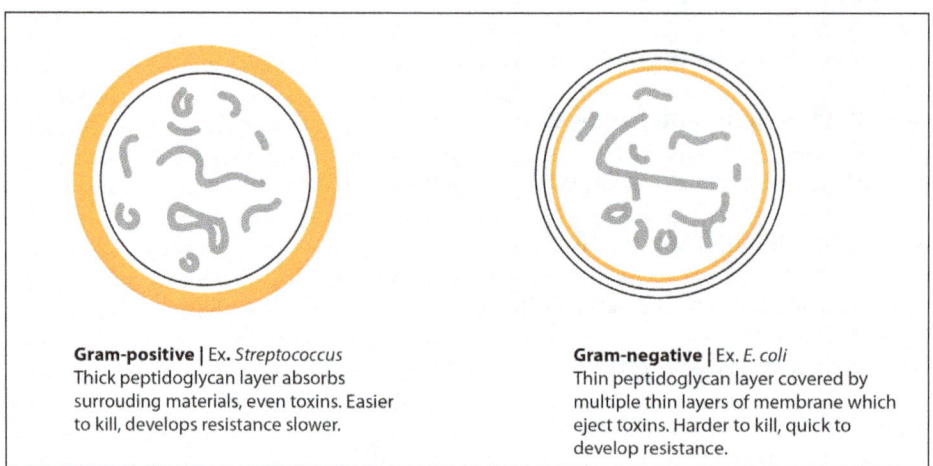

Gram-positive | Ex. *Streptococcus*
Thick peptidoglycan layer absorbs surrouding materials, even toxins. Easier to kill, develops resistance slower.

Gram-negative | Ex. *E. coli*
Thin peptidoglycan layer covered by multiple thin layers of membrane which eject toxins. Harder to kill, quick to develop resistance.

Gram-negative bacteria cannot survive as long as Gram-positive bacteria on dry surfaces (while both survive a surprisingly long time). This makes certain species more dangerous between routine cleaning, since they can survive and even multiply on dry surfaces. However, the long survival time of many pathogens means hospitals must use novel technologies to eradicate bacteria between routine cleanings.

Finally, Gram-negative bacteria are more intrinsically resistant to antibiotics - they don't absorb the toxin into their insides. Their ability to resist traditional antibiotics make them more dangerous in hospital settings, where patients are weaker and bacteria are stronger. New and very expensive antibiotics have been developed to combat these resistant species, but there remain some superbugs (MDROs) that nothing can kill. Not only do the Gram-negative bacteria's natural defences keep out these antibiotics, some even have an acquired resistance to antibiotics that make it to their inner cell bodies.

Gram-positive and Gram-negative bacteria exist everywhere, but pose unique threats to hospitalized patients with weak immune systems. Gram-positive bacteria cause tremendous problems and are the focus of many eradication efforts, but meanwhile, Gram-negative bacteria have been developing dangerous resistance and are therefore classified by the CDC as a more serious threat. For this reason, the need for new technologies that kill bacteria, both Gram-positive and Gram-negative, are essential to make hospitals safer for everyone.

References

- What-does-gram-positive-mean: fermup.com, Retrieved 28 April, 2019
- "Salmonella". Ncbi taxonomy. Bethesda, md: national center for biotechnology information. Retrieved 27 january 2019

- Gram-negative-bacteria, life-sciences: news-medical.net, Retrieved 1 August, 2019

- Fàbrega a, vila j (april 2013). "salmonella enterica serovar typhimurium skills to succeed in the host: virulence and regulation". Clinical microbiology reviews. 26 (2): 308–41. Doi:10.1128/cmr.00066-12. Pmc 3623383. Pmid 23554419

- Ryan i kj, ray cg, eds. (2004). Sherris medical microbiology (4th ed.). Mcgraw hill. Pp. 362–8. Isbn 978-0-8385-8529-0

- Proteobacteria: microscopemaster.com, Retrieved 13 July, 2019

- Davis, b; waldor, m. K. (february 2003). "filamentous phages linked to virulence of vibrio cholerae". Current opinion in microbiology. 6 (1): 35 42. Doi:10.1016/s1369-5274(02)00005-x. Pmid 12615217

- Purple-bacteria, biotech: cronodon.com, Retrieved 17 February, 2019

- Fish dn (february 2002). "optimal antimicrobial therapy for sepsis". Am j health syst pharm. 59 (suppl 1): s13–9. Doi:10.1093/ajhp/59.suppl_1.s13. Pmid 11885408

- Gram-positive-bacteria, chapter, microbiology: lumenlearning.com, Retrieved 21 April, 2019

- Salyers, abigail a. & whitt, dixie d. (2002). Bacterial pathogenesis: a molecular approach, 2nd ed. Washington, d.c.: asm press. Isbn 978-1-55581-171-6

- Gram-positive-vs-gram-positive: eoscu.com, Retrieved 28 May, 2019

- "Todar's online textbook of bacteriology: staphylococcus aureus and staphylococcal disease". Kenneth todar, phd. Retrieved dec 7, 2013

4

Bacterial Biology

The shape, projections and cytoplasm of bacteria are studied within the domain of bacterial biology. It is also involved in the genetics of bacteria as well as their growth and multiplication. These diverse aspects of bacterial biology as well as the interactions of bacteria with other organisms have been thoroughly discussed in this chapter.

An understanding of the fundamentals of bacterial biology is critical to bacteriologists and other forensic investigators attempting to identify potential biogenic pathogens that may be exploited as agents in biological warfare or by bioterrorists.

Bacteria maintain their genetic material, deoxyribonucleic acid (DNA), in a single, circular chain. Bacteria also contain DNA in small circular molecules termed plasmids.

The Dutch merchant and amateur scientist Anton van Leeuwenhoek was the first to observe bacteria and other microorganisms. Using single-lens microscopes of his own design, he described bacteria and other microorganisms as "animacules."

In addition to not being contained in a membrane bound nucleus, the DNA of prokaryotes is not associated with the special chromosome proteins called histones, which are found in higher organisms. In addition, prokaryotic cells lack other membrane-bounded organelles, such as mitochondria.

Although all bacteria share certain structural, genetic, and metabolic characteristics, important biochemical differences exist among the many species of bacteria. The cytoplasm of all bacteria is enclosed within a cell membrane surrounded by a rigid cell wall whose polymers, with few exceptions, include peptidoglycans—large, structural molecules made of protein carbohydrate. Bacteria also secrete a viscous, gelatinous polymer (called the glycocalyx) on their cell surfaces. This polymer, composed either of polysaccharide, polypeptide, or both, is called a capsule when it occurs as an organized layer firmly attached to the cell wall. Capsules increase the disease-causing ability (virulence) of bacteria by inhibiting immune system cells called phagocytes from engulfing them. The shape of bacterial cells are classified as spherical (coccus), rodlike (bacillus), spiral (spirochete), helical (spirilla) and comma-shaped (vibrio). Many bacilli and vibrio bacteria have whiplike appendages (called flagella) protruding from the cell surface. Flagella are composed of tight, helical rotors made of chains of globular protein called flagellin, and act as tiny propellers, making the bacteria very mobile. On the surface of some bacteria are short, hairlike, proteinaceous projections that may arise at the ends of the cell or over the entire surface. These projections, called fimbriae, facilitate bacteria adherence to surfaces.

Other proteinaceous projections, called pili, occur singly or in pairs, and join pairs of bacteria together, facilitating transfer of DNA between them.

During periods of harsh environmental conditions some bacteria can produce within themselves a dehydrated, thick-walled endospore. These endospores can survive extreme temperatures, dryness, and exposure to many toxic chemicals and to radiation. Endospores can remain dormant for long periods (hundreds of years in some cases) before being reactivated by the return of favorable conditions.

Identifying and Classifying Bacteria

The identification schemes of Bergey's Manual are based on morphology (e.g., coccus, bacillus), staining (gram-positive or negative), cell wall composition (e.g., presence or absence of peptidoglycan), oxygen requirements (e.g., aerobic, facultatively anaerobic) and biochemical tests (e.g., in which sugars are aerobically metabolized or fermented).

Another important identification technique is based on the principles of antigenicity—the ability to stimulate the formation of antibodies by the immune system. Commercially available solutions of antibodies against specific bacteria (antisera) are used to identify unknown organisms in a procedure called a slide agglutination test. A sample of unknown bacteria in a drop of saline is mixed with antisera that has been raised against a known species of bacteria. If the antisera causes the unknown bacteria to clump (agglutinate), then the test positively identifies the bacteria as being identical to that against which the anti-sera was raised. The test can also be used to distinguish between strains, slightly different bacteria belonging to the same species.

Pathogens are disease-causing bacteria that release toxins or poisons that interfere with some function of the host's body.

Aerobic and anaerobic bacteria. Oxygen may or may not be a requirement for a particular species of bacteria, depending on the type of metabolism used to extract energy from food (aerobic or anaerobic). Obligate aerobes must have oxygen in order to live. Facultative aerobes can exist in the absence of oxygen by using fermentation or anaerobic respiration. Anaerobic respiration and fermentation occur in the absence of oxygen, and produce substantially less ATP than aerobic respiration.

During the 1860s, the French microbiologist Louis Pasteur studied fermenting bacteria. He demonstrated that fermenting bacteria could contaminate wine and beer during manufacturing, turning the alcohol produced by yeast into acetic acid (vinegar). Pasteur also showed that heating the beer and wine to kill the bacteria preserved the flavor of these beverages. The process of heating, now called pasteurization in his honor, is still used to kill bacteria in some alcoholic beverages, as well as milk.

Pasteur described the spoilage by bacteria of alcohol during fermentation as being a "disease" of wine and beer. His work was thus vital to the later idea that human diseases could also be caused by microorganisms and that heating can destroy them.

Bacterial Growth and Division

A population of bacteria in a liquid medium is referred to as a culture. In the laboratory, where growth conditions of temperature, light intensity, and nutrients can be made ideal for the bacteria, measurements of the number of living bacteria typically reveals four stages, or phases, of growth, with respect to time. Initially, the number of bacteria in the population is low. Often the bacteria are also adapting to the environment. This represents the lag phase of growth. Depending on the health of the bacteria, the lag phase may be short or long. The latter occurs if the bacteria are damaged or have just been recovered from deep-freeze storage.

After the lag phase, the numbers of living bacteria rapidly increases. Typically, the increase is exponential. That is, the population keeps doubling in number at the same rate. This is called the log or logarithmic phase of culture growth, and is the time when the bacteria are growing and dividing at their maximum speed.

The explosive growth of bacteria cannot continue forever in the closed conditions of a flask of growth medium. Nutrients begin to become depleted, the amount of oxygen becomes reduced, and the pH changes, and toxic waste products of metabolic activity begin to accumulate. The bacteria respond to these changes in a variety of ways to do with their structure and activity of genes. With respect to bacteria numbers, the increase in the population stops and the number of living bacteria plateaus. This plateau period is called the stationary phase. Here, the number of bacteria growing and dividing is equaled by the number of bacteria that are dying.

Finally, as conditions in the culture continue to deteriorate, the proportion of the population that is dying becomes dominant. The number of living bacteria declines sharply over time in what is called the death or decline phase.

Bacteria growing as colonies on a solid growth medium also exhibit these growth phases in different regions of a colony. For example, the bacteria buried in the oldest part of the colony are often in the stationary or death phase, while the bacteria at the periphery of the colony are in the actively-dividing lo phase of growth.

Culturing of bacteria is possible such that fresh growth medium can be added at a rate equal to the rate at which culture is removed. The rate at which the bacteria grow is dependent on the rate of addition of the fresh medium. Bacteria can be tailored to grow relatively slow or fast and, if the set-up is carefully maintained, can be maintained for a long time.

Bacterial growth requires the presence of environmental factors. For example, if a bacterium uses organic carbon for energy and structure (chemoheterotrophic bacteria) then sources of carbon are needed. Such sources include simple sugars (glucose and fructose are two examples). Nitrogen is needed to make amino acids, proteins, lipids and other components. Sulphur and phosphorus are also needed for the manufacture of bacterial components. Other elements, such as potassium, calcium, magnesium, iron, manganese, cobalt and zinc are necessary for the functioning of enzymes and other processes.

Bacterial growth is also often sensitive to temperature. Depending on the species, bacteria exhibit a usually limited range in temperatures in which they can growth and reproduce. For example, bacteria known as mesophiles prefer temperatures from 20°–50 °C (68°–122 °F). Outside this range, growth and even survival is limited. Other factors, which vary depending on species, required for growth include oxygen level, pH, osmotic pressure, light and moisture.

The events of growth and division that are apparent from measurement of the numbers of living bacteria are the manifestation of a number of molecular events. At the level of the individual bacterium, the process of growth and replication is known as binary division. Binary division occurs in stages. First, the parent bacterium grows and becomes larger. Next, the genetic material inside the bacterium uncoils from the normal helical configuration and replicates. The two copies of the genetic material migrate to either end of the bacterium. Then a cross-wall known as a septum is initiated almost precisely at the middle of the bacterium. The septum grows inward as a ring from the inner surface of the membrane. When the septum is complete, an inner wall has been formed, which divides the parent bacterium into two so-called daughter bacteria. This whole process represents the generation time.

Bacteria and Disease

Bacteria can multiply and cause an infection in the bloodstream. The invasion of the bloodstream by the particular type of bacteria is referred to as a bacteremia. If the invading bacteria also release toxins into the bloodstream, the malady can also be called blood poisoning or septicemia. Staphylococcus and Streptococcus are typically associated with septicemia.

The bloodstream is susceptible to invasion by bacteria that gain entry via a wound or abrasion in the protective skin overlay of the body, or as a result of another infection elsewhere in the body, or following the introduction of bacteria during a surgical procedure or via a needle during injection of a drug.

Depending on the identity of the infecting bacterium and on the physical state of the human host (primarily with respect to the efficiency of the immune system), bacteremic infections may not produce any symptoms. However, some infections do produce symptoms, ranging from an elevated temperature, as the immune system copes with the infection, to a spread of the infection to the heart (endocarditis or pericarditis) or the covering of nerve cells (meningitis). In more rare instances, a bacteremic infection can produce a condition known as septic shock. The latter occurs when the infection overwhelms the ability of the body's defense mechanisms to cope. Septic shock can be lethal.

Septicemic infections usually result from the spread of an established infection. Bacteremic (and septicemic) infections often arise from bacteria that are normal resident on the surface of the skin or internal surfaces, such as the intestinal tract epithelial cells. In their normal environments the bacteria are harmless and even can be beneficial. However, if they gain entry to other parts of the body, these so-called commensal bacteria can pose a health threat. The entry of these commensal bacteria into the bloodstream is a normal occurrence for most people. In the majority of people, however, the immune system is more than able to deal with the invaders. If the immune system is not functioning efficiently then the invading bacteria may be able to multiply and establish an infection. Examples of conditions that compromise the immune system are another illness (such as acquired immunodeficiency syndrome and certain types of cancer), certain medical treatments such as irradiation, and the abuse of drugs or alcohol.

Examples of bacteria that are most commonly associated with bacteremic infections are Staphylococcus, Streptococcus, Pseudomonas, Haemophilus, and Escherichia coli.

The generalized location of bacteremia produces generalized symptoms. These symptoms can include a fever, chills, pain in the abdomen, nausea with vomiting, and a general feeling of ill health. Not all these symptoms are present at the same time. The nonspecific nature of the symptoms may prevent a physician from suspecting bacteremia until the infection is more firmly established. Septic shock produces more drastic symptoms, including elevated rates of breathing and heartbeat, loss of consciousness and failure of organs throughout the body. The onset of septic shock can be rapid, so prompt medical attention is critical.

As with many other infections, bacteremic infections can be prevented by observance of proper hygienic procedures including hand washing, cleaning of wounds, and cleaning sites of injections to temporarily free the surface of living bacteria. The rate of bacteremic infections due to surgery is much less now than in the past, due to the advent of sterile surgical procedures, but is still a serious concern.

Bacterial infection does not always result in disease—even if a pathogen is virulent (able to cause disease). The steps of pathogenesis (the process of causing actual disease) can depend on a number of genetic and environmental factors. In some cases, pathogenic bacteria produce toxins released extracellularly (exotoxins) that migrate from the actual site of infection to cause damage to cells in other parts of the body.

BACTERIAL GENETICS

Bacterial genetics is the subfield of genetics devoted to the study of bacteria. Bacterial genetics are subtly different from eukaryotic genetics, however bacteria still serve as a good model for animal genetic studies. One of the major distinctions between bacterial and eukaryotic genetics stems from the bacteria's lack of membrane-bound organelles (this is true of all prokaryotes. While it is a fact that there are prokaryotic organelles, they are never bound by a lipid membrane, but by a shell of proteins), necessitating protein synthesis occur in the cytoplasm.

Like other organisms, bacteria also breed true and maintain their characteristics from generation to generation, yet at the same time, exhibit variations in particular properties in a small proportion of their progeny. Though heritability and variations in bacteria had been noticed from the early days of bacteriology, it was not realised then that bacteria too obey the laws of genetics. Even the existence of a bacterial nucleus was a subject of controversy. The differences in morphology and other properties were attributed by Nageli in 1877, to bacterial pleomorphism, which postulated the existence of a single, a few species of bacteria, which possessed a protein capacity for a variation. With the development and application of precise methods of pure culture, it became apparent that different types of bacteria retained constant form and function through successive generations. This led to the concept of monomorphism.

Transformation

Transformation in bacteria was first observed in 1928 by Frederick Griffith and later (in 1944) examined at the molecular level by Oswald Avery and his colleagues who used the process to demonstrate that DNA was the genetic material of bacteria. In transformation, a cell takes up extraneous DNA found in the environment and incorporates it into its genome (genetic material) through recombination. Not all bacteria are competent to be transformed, and not all extracellular DNA is competent to transform. To be competent to transform, the extracellular DNA must be double-stranded and relatively large. To be competent to be transformed, a cell must have the surface protein Competent Factor', which binds to the extracellular DNA in an energy requiring reaction. However bacteria that are not naturally competent can be treated in such a way to make them competent, usually by treatment with calcium chloride, which make them more permeable.

Bacterial Conjugation

Bacterial conjugation is the transfer of genetic material (plasmid) between bacterial cells by direct cell-to-cell contact or by a bridge-like connection between two cells. Discovered in 1946 by Joshua Lederberg and Edward Tatum, conjugation is a mechanism of horizontal gene transfer as are transformation and transduction although these two other mechanisms do not involve cell-to-cell contact.

Bacterial conjugation is often regarded as the bacterial equivalent of sexual reproduction or mating since it involves the exchange of genetic material. During conjugation the donor cell provides a conjugative or mobilizable genetic element that is most often a plasmid or transposon. Most conjugative plasmids have systems ensuring that the recipient cell does not already contain a similar element.

The genetic information transferred is often beneficial to the recipient. Benefits may include antibiotic resistance, xenobiotic tolerance or the ability to use new metabolites. Such beneficial plasmids may be considered bacterial endosymbionts. Other elements, however, may be viewed as bacterial parasites and conjugation as a mechanism evolved by them to allow for their spread.

BACTERIA CELL

The bacteria are unicellular, achlorophyllous prokaryotic micro-organisms leading either saprophytic or parasitic mode of life. An average bacterial cell may measure 1.25 μ in diameter. The smallest among the known bacteria is dialister pneumosintes (0.15 to 0.3 μ in length) and the largest bacterium is spirillum volutans (13 to 15 μ in length).

The bacterial cell may be:

- Spherical (coccus),

- Rod shaped or cylindrical (bacillus),

- Spiral (spirillum) or spirochetes.

The spherical bacteria (cocci) may occur either singly (micro or monococci) or in pairs (diplococci) or in a group of 4 (tetracocci) or in chains (streptococci) or in irregular bunches (staphylococci). Rod shaped bacteria occur usually singly but may occasionally be found in pairs (diplobacilli) or in chains (streptobacilli).

Structure of the Bacteria Cell

The bacterial cells show a typical prokaryotic organization consisting of the following parts.

- Outer covering,

- Cytoplasm,

- Nucleoid.

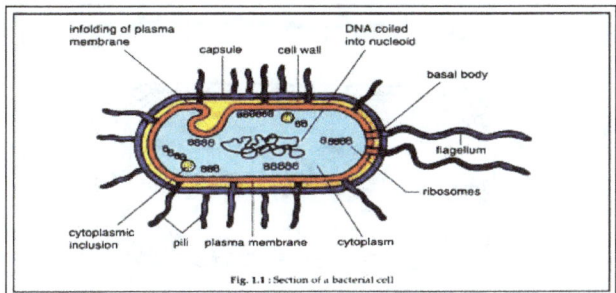

A generalized structure of bacterium.

Outer Covering of the Cell

The outer covering in most of the bacterial cells consists of the following layers:

- Capsule,

- Cell wall, and

- Plasma membrane.

Capsule

In some bacteria there may be a slimy layer outside the cell wall which is called capsule. The slimy capsule is secreted by protoplasm and is made up of di or polysaccharides. The polysaccharides of most bacterial capsules contain more than one type of sugar residues and so they are heteropolysaccharides such as D-glucose, D-galactose, D-Mannose, D-glucouronic and L-rhamnose residues etc.

In some bacteria the polysaccharides do contain single sugar residue and so they are homopolysaccharides such as polyfructose.

Capsules may be divided into two categories:

- Macrocapsules,

- Microcapsules.

Macrocapsules are about 0.2 m thick and can be seen under light microscope after special staining. Microcapsules are extremely thin and cannot be seen under light microscope. They can, however, be demonstrated immunologically. The capsule layer serves as a protective covering for bacterial protoplasm. In pathogenic bacteria capsule provides enough resistance against drugs and phagocytes.

Appreciable number of bacterial species is motile and they are provided with flagella. Bacterial flagella are about 100 to 200 A thick and show varied length. They develop from basal granules lying below the cell wall. The bacterial flagella, unlike multi-stranded flagella of eukaryotes, are made up of special type of proteins.

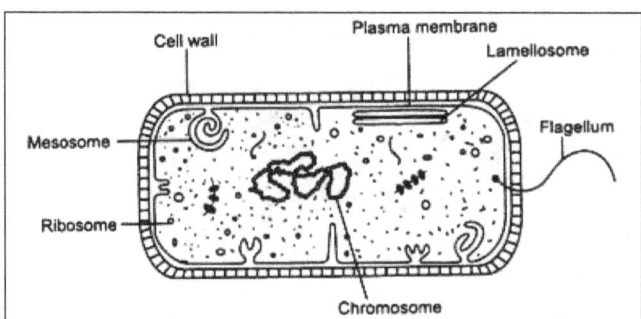

Structure of a bacterial cell.

In addition to flagella, there occur some hair-like or peg-like outgrowths on the surface of some bacterial cells. These are called pili or fimbriae. They are composed of helically arranged units of special protein called pilin. Pilin helps in the attachment of bacterial cells to some other objects, and in some they act as conjugation channels through which DNA of donor cell moves into the female cell.

Cell Wall

Cell wall is rigid structure, 10 nm thick which is present just below the capsule and outside the plasma membrane. The cell wall maintains the shape of the cell and protects the cell from high osmotic pressure gradient. The chemical composition of bacterial cell wall is different from that of higher plant cell. It does not contain cellulose.

The main structural components of almost all bacterial cell walls are peptidoglycans or mucopeptides consisting of acetylglucosamine and acetyl muramic acid molecules linked alternately in chain. In certain bacteria, Teichoic acid molecules are covalently linked to peptidoglucan.

Other compounds found in the cell wall are proteins, polysaccharides, lipids, certain inorganic salts, phosphorous and aminoacids, diaminopimallic acid (a substance found only in bacteria and blue-green algae).

The polysaccharides of bacterial cell wall consist of different sugars such as glucose, galactose and mannose or corresponding aminosugars depending upon bacteria. The electron microscopy of bacterial cell wall reveals that it is composed of granular units of 50 to 140 nm diameter arranged in regular hexagonal or rectangular pattern.

Plasma Membrane

Beneath the cell wall there lies a thin living membrane, the plasma membrane which forms the outer boundary of cytoplasm. The plasma membrane is 75 Å thick unit membrane and is composed chiefly of proteins and phospholipids.

Some amount of carbohydrate, DNA and RNA have also been reported from plasma membrane but it still needs substantial proof. The lipids in the prokaryotic plasma membrane are polar lipids which may be glycophosphates or glycolipids. Small amount of quinone, Co-enzyme-Q, vitamin K and carotenoids may also be found in the bacterial plasma membrane.

This membrane is a selective barrier to the surrounding medium. The plasma membranes of bacteria contain enzymes involved in oxidation of metabolites or respiratory chain and many multienzyme complexes. It performs the functions including oxidative phosphorylation which are usually done by mitochondria in eukaryotic cells.

Bacterial plasma membrane contains many specific transport systems for compounds of the sugars, aminoacids, mineral ions etc. The plasma membrane is capable of not only transporting materials by simple diffusion but is also involved in active transport against concentration gradient.

At certain places membrane may be infolded to form whorls of convoluted membranes called mesosomes. The mesosomes perform several important metabolic activities such as respiration, secretion, etc. They are thought to increase the surface area for transporting systems of the cells. They are also sites of DNA replication enzymes and nucleoid separation.

Cytoplasm

The cytoplasm of bacterial cell is dense and colloidal substance containing a variety of organic compounds such as proteins, glycogens and lipids. Besides, granulose (a polymer of glucose), volutin (polymetaphosphate), polybeta-hydroxibutyric acid and elemental sulphur may also be found in the form of granules.

The majority of cytoplasmic organelles such as endoplasmic reticulum, golgi complex, plastids, mitochondria, lysosomes and centrioles are lacking in the bacterial cytoplasm. Photosynthetic bacteria contain a pigment called bacteriochlorophyll which is somewhat different from the chlorophylls found in eukaryotes.

The photosynthetic pigments and enzymes are generally found in the groundplasm associated with the lamellae, tubules or vesicles called chromatophores. The bacterial cells contain ribosome particles of about 25 nm diameter which exist in the cytoplasm in free state or in the form of polysomes but are not attached to the membrane.

Ribosome particles constitute upto 30% of the bacterial weight and they are the sites for protein synthesis.

Nucleoid

Nuclear region or nucleoid contains genophore which is single circular double stranded molecule of deoxyribonucleic acid (DNA). DNA molecule of bacterium is about 1mm long (lO6nm) and contains all the genetic information.

One molecule is sufficient to code for about 2000 to 3000 different proteins. DNA molecule is folded and packed within the nuclear region. It lies free in the cytoplasm and is not bounded by nuclear membrane.

DNA molecule of bacterial cell appears to be free and not complexed with histone but contains polyamines (non-histone type) which may be linked to some of its phosphate groups. Under certain conditions bacterial cell may contain two or more DNA molecules, because of replication of original DNA.

Evidence from studies of Bacillus subtitles indicates that the circular DNA may be attached to the plasma membrane via mesosome. So, it is possible that mesosome plays an important role in DNA replication by providing a mechanism for unwinding of DNA double helix as well as energy. In general, during division of the bacterial cell DNA molecule replicates and the two molecules go to different daughter cells.

Bacterial Cell Structure

The bacterium, despite its simplicity, contains a well-developed cell structure which is responsible for some of its unique biological structures and pathogenicity. Many structural features are unique to bacteria and are not found among archaea or eukaryotes. Because of the simplicity of bacteria relative to larger organisms and the ease with which they can be manipulated experimentally, the cell structure of bacteria has been well studied, revealing many biochemical principles that have been subsequently applied to other organisms.

Cell Morphology

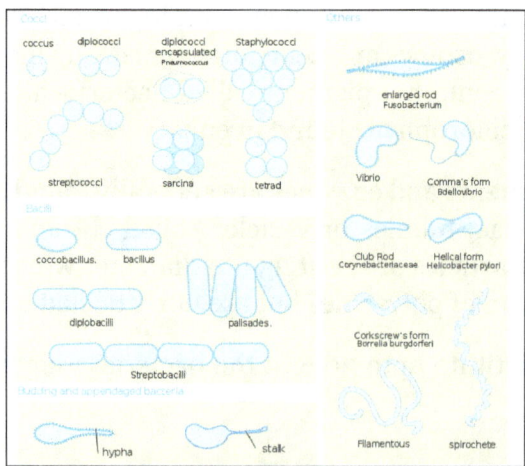

Bacteria come in a wide variety of shapes

Perhaps the most elemental structural property of bacteria is their morphology (shape). Typical examples include:

- Coccus (circle or spherical),

- Bacillus (rod-like),

- Coccobacillus (between a sphere and a rod),

- Spiral (corkscrew-like),

- Filamentous (elongated).

Cell shape is generally characteristic of a given bacterial species, but can very depending on growth conditions. Some bacteria have complex life cycles involving the production of stalks and appendages (e.g. *Caulobacter*) and some produce elaborate structures bearing reproductive spores (e.g. *Myxococcus, Streptomyces*). Bacteria generally form distinctive cell morphologies when examined by light microscopy and distinct colony morphologies when grown on Petri plates.

Perhaps the most obvious structural characteristic of bacteria is (with some exceptions) their small size. For example, *Escherichia coli* cells, an "average" sized bacterium, are about 2 μm (micrometres) long and 0.5 μm in diameter, with a cell volume of 0.6–0.7 μm^3. This corresponds to a wet mass of about 1 picogram (pg), assuming that the cell consists mostly of water. The dry mass of a single cell can be estimated as 23% of the wet mass, amounting to 0.2 pg. About half of the dry mass of a bacterial cell consists of carbon, and also about half of it can be attributed to proteins. Therefore, a typical fully grown 1-liter culture of *Escherichia coli* (at an optical density of 1.0, corresponding to c. 10^9 cells/ml) yields about 1 g wet cell mass. Small size is extremely important because it allows for a large surface area-to-volume ratio which allows for rapid uptake and intra-cellular distribution of nutrients and excretion of wastes. At low surface area-to-volume ratios the diffusion of nutrients and waste products across the bacterial cell membrane limits the rate at which microbial metabolism can occur, making the cell less evolution-arily fit. The reason for the existence of large cells is unknown, although it is speculated that the increased cell volume is used primarily for storage of excess nutrients.

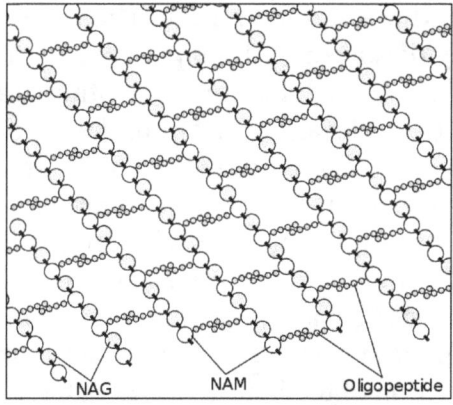

The structure of peptidoglycan.

The cell envelope is composed of the plasma membrane and cell wall. As in other organisms, the bacterial cell wall provides structural integrity to the cell. In prokaryotes, the primary function of the cell wall is to protect the cell from internal turgor pressure caused by the much higher concentrations of proteins and other molecules inside the cell compared to its external environment. The bacterial cell wall differs from that of all other organisms by the presence of peptidoglycan which is located immediately outside of the cytoplasmic membrane. Peptidoglycan is made up of a polysaccharide backbone consisting of alternating N-Acetylmuramic acid (NAM) and N-acetylglucos-amine (NAG) residues in equal amounts. Peptidoglycan is responsible for the rigidity of the bacterial cell wall and for the determination of cell shape. It is relatively porous and is not considered to be a permeability barrier for small substrates. While all bacterial cell walls (with a few exceptions e.g. extracellular parasites such as Mycoplasma) contain peptidoglycan, not all cell walls have the same overall structures. Since the cell wall is required for bacterial survival, but is absent in some eukaryotes, several antibiotics (notably the penicillins and cephalosporins) stop bacterial infections by interfering with cell wall synthesis, while having no effects on human cells which have no cell wall, only a cell membrane. There are two main types of bacterial cell walls, those of gram-positive bacteria and those of gram-negative bacteria, which are differentiated by their Gram staining characteristics. For both these types of bacteria, particles of approximately 2 nm can pass through the peptidoglycan. If the bacterial cell wall is entirely removed, it is called a protoplast while if it's partially removed, it is called a spheroplast. β-Lactam antibiotics such as penicillin inhibit the formation of peptidoglycan cross-links in the bacterial cell wall. The enzyme lysozyme, found in human tears, also digests the cell wall of bacteria and is the body's main defense against eye infections.

The Gram positive Cell Wall

Gram-positive cell walls are thick and the peptidoglycan (also known as murein) layer constitutes almost 95% of the cell wall in some gram-positive bacteria and as little as 5-10% of the cell wall in gram-negative bacteria. The gram-positive bacteria take up the crystal violet dye and are stained purple. The cell wall of some gram-positive bacteria can be completely dissolved by lysozymes which attacks the bonds between N-acetylmuramic acid and N-acetylglucosamine. In other gram-positive bacteria, such as Staphylococcus aureus, the walls are resistant to the action of lysozymes. They have O-acetyl groups on carbon-6 of some muramic acid residues. The matrix substances in the walls of gram-positive bacteria may be polysaccharides or teichoic acids. The latter are very widespread, but have been found only in gram-positive bacteria. There are two main types of teichoic acid: ribitol teichoic acids and glycerol teichoic acids. The latter one is more widespread. These acids are polymers of ribitol phosphate and glycerol phosphate, respectively, and only located on the surface of many gram-positive bacteria. However, the exact function of teichoic acid is debated and not fully understood. A major component of the gram-positive cell wall is lipoteichoic acid. One of its purposes is providing an antigenic function. The lipid element is to be found in the membrane where its adhesive properties assist in its anchoring to the membrane.

Gram Negative Cell Wall

Gram-negative cell walls are thin and unlike the gram-positive cell walls, they contain a thin peptidoglycan layer adjacent to the cytoplasmic membrane. Gram-negative bacteria are stained as pink colour. The chemical structure of the outer membrane's lipopolysaccharide is often unique to specific bacterial sub-species and is responsible for many of the antigenic properties of these strains.

Plasma Membrane

The plasma membrane or bacterial cytoplasmic membrane is composed of a phospholipid bilayer and thus has all of the general functions of a cell membrane such as acting as a permeability barrier for most molecules and serving as the location for the transport of molecules into the cell. In addition to these functions, prokaryotic membranes also function in energy conservation as the location about which a proton motive force is generated. Unlike eukaryotes, bacterial membranes (with some exceptions e.g. Mycoplasma and methanotrophs) generally do not contain sterols. However, many microbes do contain structurally related compounds called hopanoids which likely fulfill the same function. Unlike eukaryotes, bacteria can have a wide variety of fatty acids within their membranes. Along with typical saturated and unsaturated fatty acids, bacteria can contain fatty acids with additional methyl, hydroxy or even cyclic groups. The relative proportions of these fatty acids can be modulated by the bacterium to maintain the optimum fluidity of the membrane (e.g. following temperature change).

As a phospholipid bilayer, the lipid portion of the outer membrane is impermeable to charged molecules. However, channels called porins are present in the outer membrane that allow for passive transport of many ions, sugars and amino acids across the outer membrane. These molecules are therefore present in the periplasm, the region between the cytoplasmic and outer membranes. The periplasm contains the peptidoglycan layer and many proteins responsible for substrate binding or hydrolysis and reception of extracellular signals. The periplasm is thought to exist in a gel-like state rather than a liquid due to the high concentration of proteins and peptidoglycan found within it. Because of its location between the cytoplasmic and outer membranes, signals received and substrates bound are available to be transported across the cytoplasmic membrane using transport and signalling proteins imbedded there.

Extracellular (External) Structures

Fimbriae and Pili

Fimbriae (sometimes called "attachment pili") are protein tubes that extend out from the outer membrane in many members of the Proteobacteria. They are generally short in length and present in high numbers about the entire bacterial cell surface. Fimbriae usually function to facilitate the attachment of a bacterium to a surface (e.g. to form a biofilm) or to other cells (e.g. animal cells during pathogenesis). A few organisms (e.g. Myxococcus) use fimbriae for motility to facilitate the assembly of multicellular structures such as fruiting bodies. Pili are similar in structure to fimbriae but are much longer and present on the bacterial cell in low numbers. Pili are involved in the process of bacterial conjugation where they are called conjugation pili or "sex pili". Type IV pili (non-sex pili) also aid bacteria in gripping surfaces.

S-layers

An S-layer (surface layer) is a cell surface protein layer found in many different bacteria and in some archaea, where it serves as the cell wall. All S-layers are made up of a two-dimensional array of proteins and have a crystalline appearance, the symmetry of which differs between species. The exact function of S-layers is unknown, but it has been suggested that they act as a partial permeability barrier for large substrates. For example, an S-layer could conceivably keep extracellular proteins near the cell membrane by preventing their diffusion away from the cell. In some pathogenic species, an S-layer may help to facilitate survival within the host by conferring protection against host defence mechanisms.

Glycocalyx

Many bacteria secrete extracellular polymers outside of their cell walls called glycocalyx. These polymers are usually composed of polysaccharides and sometimes protein. Capsules are relatively impermeable structures that cannot be stained with dyes such as India ink. They are structures that help protect bacteria from phagocytosis and desiccation. Slime layer is involved in attachment of bacteria to other cells or inanimate surfaces to form biofilms. Slime layers can also be used as a food reserve for the cell.

Flagella

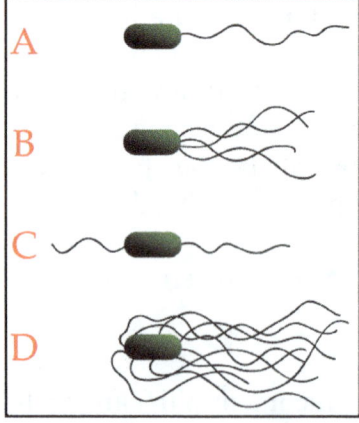

A-Monotrichous; B-Lophotrichous; C-Amphitrichous; D-Peritrichous.

Perhaps the most recognizable extracellular bacterial cell structures are flagella. Flagella are whip-like structures protruding from the bacterial cell wall and are responsible for bacterial motility (i.e. movement). The arrangement of flagella about the bacterial cell is unique to the species observed. Common forms include:

- Monotrichous – Single flagellum.

- Lophotrichous – A tuft of flagella found at one of the cell poles.

- Amphitrichous – Single flagellum found at each of two opposite poles.

- Peritrichous – Multiple flagella found at several locations about the cell.

The bacterial flagellum consists of three basic components: a whip-like filament, a motor complex, and a hook that connects them. The filament is approximately 20 nm in diameter and consists of several protofilaments, each made up of thousands of flagellin subunits. The bundle is held together by a cap and may or may not be encapsulated. The motor complex consists of a series of rings anchoring the flagellum in the inner and outer membranes, followed by a proton-driven motor that drives rotational movement in the filament.

Intracellular (Internal) Structures

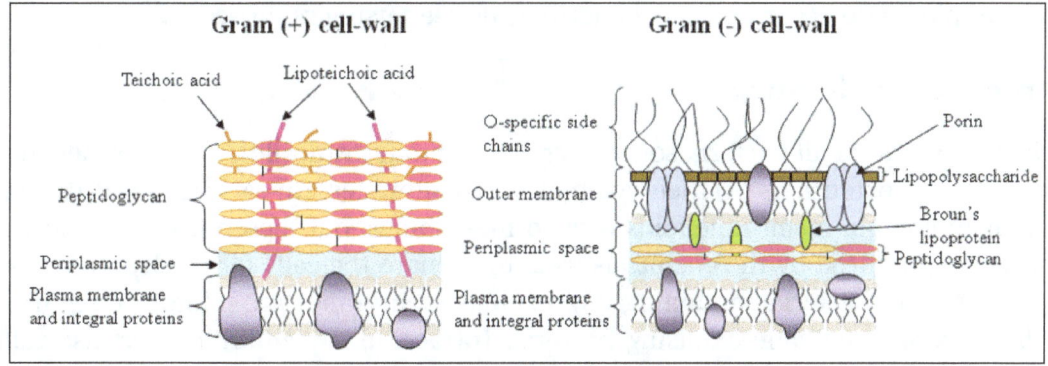

Cell structure of a gram positive bacterium.

In comparison to eukaryotes, the intracellular features of the bacterial cell are extremely simple. Bacteria do not contain organelles in the same sense as eukaryotes. Instead, the chromosome and perhaps ribosomes are the only easily observable intracellular structures found in all bacteria. They do exist, however, specialized groups of bacteria that contain more complex intracellular structures.

Bacterial DNA and Plasmids

Unlike eukaryotes, the bacterial DNA is not enclosed inside of a membrane-bound nucleus but instead resides inside the bacterial cytoplasm. This means that the transfer of cellular information through the processes of translation, transcription and DNA replication all occur within the same compartment and can interact with other cytoplasmic structures, most notably ribosomes. The bacterial DNA is not packaged using histones to form chromatin as in eukaryotes but instead exists as a highly compact supercoiled structure, the precise nature of which remains unclear. Most bacterial chromosomes are circular although some examples of linear DNA exist (e.g. Borrelia burgdorferi). Along with chromosomal DNA, most bacteria also contain small independent pieces of DNA called plasmids that often encode for traits that are advantageous but not essential to their bacterial host. Plasmids can be easily gained or lost by a bacterium and can be transferred between bacteria as a form of horizontal gene transfer. So plasmids can be described as an extra chromosomal DNA in a bacterial cell.

Ribosomes and other Multiprotein Complexes

In most bacteria the most numerous intracellular structure is the ribosome, the site of protein synthesis in all living organisms. All prokaryotes have 70S (where S=Svedberg units) ribosomes while eukaryotes contain larger 80S ribosomes in their cytosol. The 70S ribosome is made up of a 50S and 30S subunits. The 50S subunit contains the 23S and 5S rRNA while the 30S subunit contains the 16S rRNA. These rRNA molecules differ in size in eukaryotes and are complexed with a large number of ribosomal proteins, the number and type of which can vary slightly between organisms. While the ribosome is the most commonly observed intracellular multiprotein complex in bacteria other large complexes do occur and can sometimes be seen using microscopy.

Intracellular Membranes

While not typical of all bacteria some microbes contain intracellular membranes in addition to (or as extensions of) their cytoplasmic membranes. An early idea was that bacteria might contain membrane folds termed mesosomes, but these were later shown to be artifacts produced by the chemicals used to prepare the cells for electron microscopy. Examples of bacteria containing intracellular membranes are phototrophs, nitrifying bacteria and methane-oxidising bacteria. Intracellular membranes are also found in bacteria belonging to the poorly studied Planctomycetes group, although these membranes more closely resemble organellar membranes in eukaryotes and are currently of unknown function. Chromatophores are intracellular membranes found in phototrophic bacteria. Used primarily for photosynthesis, they contain bacterio chlorophyll pigments and carotenoids.

Cytoskeleton

The prokaryotic cytoskeleton is the collective name for all structural filaments in prokaryotes. It was once thought that prokaryotic cells did not possess cytoskeletons, but recent advances in visualization technology and structure determination have shown that filaments indeed exist in these cells. In fact, homologues for all major cytoskeletal proteins in eukaryotes have been found in prokaryotes. Cytoskeletal elements play essential roles in cell division, protection, shape determination, and polarity determination in various prokaryotes.

Nutrient Storage Structures

Most bacteria do not live in environments that contain large amounts of nutrients at all times. To accommodate these transient levels of nutrients bacteria contain several different methods of nutrient storage in times of plenty for use in times of want. For example, many bacteria store excess carbon in the form of polyhydroxyalkanoates or glycogen. Some microbes store soluble nutrients such as nitrate in vacuoles. Sulfur is most often stored as elemental (S^o) granules which can be deposited either intra- or extracellularly. Sulfur granules are especially common in bacteria that use hydrogen sulfide as an electron source. Most of the above-mentioned examples can be viewed using a microscope and are surrounded by a thin nonunit membrane to separate them from the cytoplasm.

Inclusions

Inclusions are considered to be nonliving components of the cell that do not possess metabolic activity and are not bounded by membranes. The most common inclusions are glycogen, lipid droplets, crystals, and pigments. Volutin granules are cytoplasmic inclusions of complexed inorganic polyphosphate. These granules are called metachromatic granules due to their displaying the metachromatic effect; they appear red or blue when stained with the blue dyes methylene blue or toluidine blue.

Gas Vacuoles

Gas vacuoles are membrane-bound, spindle-shaped vesicles, found in some planktonic bacteria and Cyanobacteria, that provides buoyancy to these cells by decreasing their overall cell density. Positive buoyancy is needed to keep the cells in the upper reaches of the water column, so that they can continue to perform photosynthesis. They are made up of a shell of protein that has a highly hydrophobic inner surface, making it impermeable to water (and stopping water vapour from condensing inside) but permeable to most gases. Because the gas vesicle is a hollow cylinder, it is liable to collapse when the surrounding pressure increases. Natural selection has fine tuned the structure of the gas vesicle to maximise its resistance to buckling, including an external strengthening protein, GvpC, rather like the green thread in a braided hosepipe. There is a simple relationship between the diameter of the gas vesicle and pressure at which it will collapse – the wider the gas vesicle the weaker it becomes. However, wider gas vesicles are more efficient, providing more buoyancy per unit of protein than narrow gas vesicles. Different species produce gas vesicle of different diameter, allowing them to colonise different depths of the water column (fast growing, highly competitive species with wide gas vesicles in the top most layers; slow growing, dark-adapted, species with strong narrow gas vesicles in the deeper layers). The diameter of the gas vesicle will also help determine which species survive in different bodies of water. Deep lakes that experience winter mixing expose the cells to the hydrostatic pressure generated by the full water column. This will select for species with narrower, stronger gas vesicles.

The cell achieves its height in the water column by synthesising gas vesicles. As the cell rises up, it is able to increase its carbohydrate load through increased photosynthesis. Too high and the cell will suffer photobleaching and possible death, however, the carbohydrate produced during photosynthesis increases the cell's density, causing it to sink. The daily cycle of carbohydrate build-up from photosynthesis and carbohydrate catabolism during dark hours is enough to fine-tune the cell's position in the water column, bring it up toward the surface when its carbohydrate levels are low and it needs to photosynthesis, and allowing it to sink away from the harmful UV radiation when the cell's carbohydrate levels have been replenished. An extreme excess of carbohydrate causes a significant change in the internal pressure of the cell, which causes the gas vesicles to buckle and collapse and the cell to sink out.

Microcompartments

Bacterial microcompartments are widespread, membrane-bound organelles that are made of a protein shell that surrounds and encloses various enzymes. provide a further level of organization; they are compartments within bacteria that are surrounded by polyhedral protein shells, rather than by lipid membranes. These "polyhedral organelles" localize and compartmentalize bacterial metabolism, a function performed by the membrane-bound organelles in eukaryotes.

Carboxysomes

Carboxysomes are bacterial microcompartments found in many autotrophic bacteria such as Cyanobacteria, Knallgasbacteria, Nitroso- and Nitrobacteria. They are proteinaceous structures resembling phage heads in their morphology and contain the enzymes of carbon dioxide fixation in these organisms (especially ribulose bisphosphate carboxylase/oxygenase, RuBisCO, and carbonic anhydrase). It is thought that the high local concentration of the enzymes along with the fast conversion of bicarbonate to carbon dioxide by carbonic anhydrase allows faster and more efficient carbon dioxide fixation than possible inside the cytoplasm. Similar structures are known to harbor the coenzyme B12-containing glycerol dehydratase, the key enzyme of glycerol fermentation to 1,3-propanediol, in some Enterobacteriaceae (e. g. Salmonella).

Magnetosomes

Magnetosomes are bacterial microcompartments found in magnetotactic bacteria that allow them to sense and align themselves along a magnetic field (magnetotaxis). The ecological role of magnetotaxis is unknown but is thought to be involved in the determination of optimal oxygen concentrations. Magnetosomes are composed of the mineral magnetite or greigite and are surrounded by a lipid bilayer membrane. The morphology of magnetosomes is species-specific.

Endospores

Perhaps the best known bacterial adaptation to stress is the formation of endospores. Endospores are bacterial survival structures that are highly resistant to many different types of chemical and environmental stresses and therefore enable the survival of bacteria in environments that would be lethal for these cells in their normal vegetative form. It has been proposed that endospore formation has allowed for the survival of some bacteria for hundreds of millions of years (e.g. in salt crystals) although these publications have been questioned. Endospore formation is limited to several genera of gram-positive bacteria such as Bacillus and Clostridium. It differs from reproductive spores in that only one spore is formed per cell resulting in no net gain in cell number upon endospore germination. The location of an endospore within a cell is species-specific and can be used to determine the identity of a bacterium. Dipicolinic acid is a chemical compound which composes 5% to 15% of the dry weight of bacterial spores and is implicated in being responsible for the heat resistance of endospores.

The Bacterial Cell Envelope

The bacteria cell envelope is a complex multilayered structure that serves to protect these organisms from their unpredictable and often hostile environment. The cell envelopes of most bacteria fall into one of two major groups. Gram-negative bacteria are surrounded by a thin peptidoglycan cell wall, which itself is surrounded by an outer membrane containing lipopolysaccharide. Gram-positive bacteria lack an outer membrane but are surrounded by layers of peptidoglycan many times thicker than is found in the Gram-negatives. Threading through these layers of peptidoglycan are long anionic polymers, called teichoic acids.

It has been well known since the late 1830s that all living organisms are composed of fundamental units called cells. The cell is a finite entity with a definite boundary, the plasma membrane. That means that the essence of the living state must be contained within the biological membrane; it is a defining feature of all living things. Everything that exists outside of the biological membrane is nonliving. Chemists tend to think of membranes as self assembling but biological membranes do not self assemble; they require energy to be established and maintained. In almost all cells this structure is a phospholipid bilayer that surrounds and contains the cytoplasm. In addition to lipid components biological membranes are composed of proteins; the proteins are what make each membrane unique. Despite its obvious importance, membranes and their associated functions remained poorly understood until the 1950s. Before this, many viewed the membrane as a semipermeable bag. This view persisted for a number of years, particularly in the case of bacteria, because it could not be envisioned how such a "simple" organism could have anything but a simple membrane.

The bacterial cell envelope, i.e., the membrane(s) and other structures that surround and protect the cytoplasm, however, is anything but a simple membrane. Unlike cells of higher organisms, the bacterium is faced with an unpredictable, dilute and often hostile environment. To survive, bacteria have evolved a sophisticated and complex cell envelope that protects them, but allows selective passage of nutrients from the outside and waste products from the inside. It is easily appreciated that a living system cannot do what it does without the ability to establish separate compartments in which components are segregated. Specialized functions occur within different compartments because the types of molecules within the compartment can be restricted. However, membranes do not simply serve to segregate different types of molecules. They also function as surfaces on which reactions can occur. Recent advances in microscopy, have revealed strikingly nonrandom localization of envelope components.

Christian Gram developed a staining procedure that allowed him to classify nearly all bacteria into two large groups, and this eponymous stain is still in widespread use. One group of bacteria retain Christian's stain, Gram-positive, and the other do not, Gram-negative. The basis for the Gram stain lies in fundamental structural differences in the cell envelope of these two groups of bacteria. *B. subtilis* is the major Gram-positive model organism and a substantial knowledge base exists for it, but the cell envelope of *S. aureus* has been studied more extensively because of interest in how surface features mediate interactions with the environment in the course of infection. Care should be taken in generalizing from examples drawn from particular microorganisms. For example, *E. coli* inhabits the mammalian gut. Accordingly, *E. coli* and other enteric bacteria must have a cell envelope that is particularly effective at excluding detergents such as bile salts. This need not be a pressing issue for other Gram-negative bacteria, and their envelopes may differ in species- and environmentally specific ways. Nonetheless, the ability to use the Gram stain to categorize bacteria suggests that the basic organizational principles we present are conserved. In addition, many bacteria express an outermost coat, the S-layer, which is composed of a single protein that totally encases the organism. S-layers and capsules.

The Gram negative Cell Envelope

After more than a decade of controversy, techniques of electron microscopy were improved to the point in which they finally revealed a clearly layered structure of the Gram-negative cell envelope. There are three principal layers in the envelope; the outer membrane (OM), the peptidoglycan cell wall, and the cytoplasmic or inner membrane (IM). The two concentric membrane layers delimit an aqueous cellular compartment that first termed the periplasm. During a similar time frame biochemical methods were developed to isolate and characterize the distinct set of proteins found in the periplasm, and to characterize the composition of both the inner and outer membranes. Studies since then have only reinforced their basic conclusions.

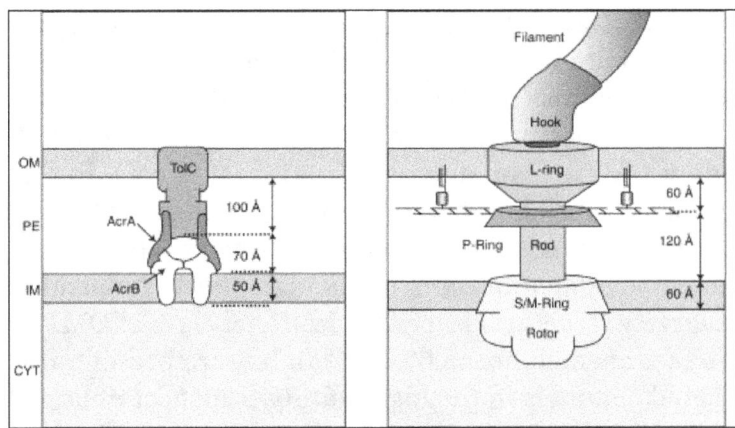

Transenvelope machines in the Gram-negative cell envelope. The AcrA/B proteins together with TolC form an efflux pump that expels harmful molecules such as antibiotics from the cell directly into the media. The flagellar basal body hook structure connects the motor to the flagella. Distances shown provide a reasonable estimate of the size of the cellular compartments shown. PE, periplasm; CYT, cytoplasm.

The Outer Membrane

Starting from the outside and proceeding inward the first layer encountered is the OM. The OM is a distinguishing feature of Gram-negative bacteria; Gram-positive bacteria lack this organelle. Like other biological membranes, the OM is a lipid bilayer, but importantly, it is not a phospholipid bilayer. The OM does contain phospholipids; they are confined to the inner leaflet of this membrane. The outer leaflet of the OM is composed of glycolipids, principally lipopolysaccharide (LPS). LPS is an infamous molecule because it is responsible for the endotoxic shock associated with the septicemia caused by Gram-negative organisms. The human innate immune system is sensitized to this molecule because it is a sure indicator of infection.

With few exceptions, the proteins of the OM can be divided into two classes, lipoproteins and β-barrel proteins. Lipoproteins contain lipid moieties that are attached to an amino-terminal cysteine residue. It is generally thought that these lipid moieties embed lipoproteins in the inner leaflet of the OM. In other words, these proteins are not thought to be transmembrane proteins. There are about 100 OM lipoproteins in *E.coli*, and the functions of most of these are not known. Nearly all of the integral, transmembrane proteins of the outer membrane assume a β-barrel conformation. These proteins are β sheets that are wrapped into cylinders. Not surprisingly, some of these OMPs, such as the porins, OmpF, and OmpC, function to allow the passive diffusion of small molecules such as mono- and disaccharides and amino acids across the OM. These porins have 16 transmembrane β strands, they exist as trimers, and they are very abundant; together they are present at approximately 250,000 copies per cell. Other OMPs, such as LamB (18 transmembrane β strands) or PhoE

(16 transmembrane β strainds), exist as trimers as well and they function in the diffusion of specific small molecules, maltose or maltodextrins and anions such as phosphate respectively, across the OM. When induced by the presence of maltose or phosphate starvation, respectively, these proteins are very abundant as well. OmpA is another abundant OMP. It is monomeric, and it is unusual in that it can exist in two different conformations. A minor form of the protein, with an unknown number of transmembrane strands, can function as a porin, but the major, nonporin form has only eight transmembrane strands, and the periplasmic domain of this form performs a largely structural role. An additional class of OMPs, which are larger β-barrels (20–24 transmembrane β strands), but are present at much lower levels, function as gated channels in the high affinity transport of large ligands such as Fe-chelates or vitamins such as vitamin B-12.

The OM is essential for the survival of *E. coli*, but it contains only a few enzymes. For example, there is a phospholipase (PldA), a protease (OmpT), and an enzyme that modifies LPS (PagP). The active site of all of these enzymes is located in the outer leaflet, or it faces the exterior of the cell (OmpT). Mutants lacking any of these enzymes exhibit no striking phenotypes. The only known function of the OM is to serve as a protective barrier, and it is not immediately obvious why this organelle is essential. But what a barrier it is. *Salmonella*, another enteric bacterium, can live at the site of bile salt production in the gall bladder, and it is generally true that Gram-negative bacteria are more resistant to antibiotics than are their Gram-positive cousins. Indeed, some Gram-negative bacteria, such as *Pseudomonas*, are notorious in this regard.

LPS plays a critical role in the barrier function of the OM. It is a glucosamine disaccharide with six or seven acyl chains, a polysaccharide core, and an extended polysaccharide chain that is called the O-antigen. Traditionally pathogenic *E. coli* are classified by the antigenic properties of their O-antigen and the major protein (flagellin, termed H) component of the flagella. Hence, *E. coli* O157:H7. LPS molecules bind each other avidly, especially if cations like Mg^{++} are present to neutralize the negative charge of phosphate groups present on the molecule. The acyl chains are largely saturated, and this facilitates tight packing. The nonfluid continuum formed by the LPS molecules is a very effective barrier for hydrophobic molecules. This coupled with the fact that the porins limit diffusion of hydrophilic molecules larger than about 700 Daltons, make the OM a very effective yet, selective permeability barrier.

The Peptidoglycan Cell Wall

Bacteria do not lyse when put into distilled water because they have a rigid exoskeleton. Peptidoglycan is made up of repeating units of the disaccharide N-acetyl glucosamine-N-actyl muramic acid, which are cross-linked by pentapeptide side chains. The peptidoglycan sacculus is one very large polymer that can be isolated and viewed in a light microscope. Because of its rigidity, it determines cell shape. The enterics are rod shaped, but cell shapes can vary. For example, vibrios and caulobacters are comma shaped. Recent results suggest that the glycan chains run perpendicular to the long axis of a rod shaped cell, i.e., hoops of glycan chains around the girth of the cell. Agents such as enzymes or antibiotics that damage the peptidoglycan cause cell lysis owing to the turgor pressure of the cytoplasm. Lysis can be prevented in media of high osmolarity. However, without the peptidoglycan, cells lose their characteristic shape. The resulting cells are called spheroplasts. With *E. coli*, normal methods of spheroplast production produce nonviable cells, but they can continue metabolism and biosynthesis for hours. However, special methods can be used to produce L forms, which are spherical in shape and can be propagated on high osmolarity media.

The OM is basically stapled to the underlying peptidoglycan by a lipoprotein called Lpp, murein lipoprotein, or Braun's lipoprotein. The lipids attached to the amino terminus of this small protein (58 amino acids) embed it in the OM. Lpp is the most abundant protein in *E. coli*, more than 500,000 molecules per cell. The ε-amino group of the carboxy-terminal lysine residue of one third of these molecules is covalently attached to the diaminopimelate residue in the peptide crossbridge. In addition, proteins such as OmpA bind peptidoglycan noncovalently. Nonetheless, mutants that lack Lpp shed OM vesicles.

The Periplasm

The OM and IM delimit an aqueous cellular compartment called the periplasm. The periplasm is densely packed with proteins and it is more viscous than the cytoplasm. Cellular compartmentalization allows Gram-negative bacteria to sequester potentially harmful degradative enzymes such as RNAse or alkaline phosphatase. Because of this, the periplasm has been called an evolutionary precursor of the lysosomes of eukaryotic cells. Other proteins that inhabit this compartment include the periplasmic binding proteins, which function in sugar and amino acid transport and chemotaxis, and chaperone-like molecules that function in envelope biogenesis.

The Inner Membrane

One of the hallmarks of eukaryotic cells is the presence of intracellular organelles. These organelles are defined by limiting membranes, and these organelles perform a number of essential cellular processes. The mitochondria produce energy, the smooth endoplasmic reticulum (ER) synthesize lipids, protein secretion occurs in the rough ER, and the cytoplasmic membrane contains the receptors that sense the environment and the transport systems for nutrients and waste products. Bacteria lack intracellular organelles, and consequently, all of the membrane-associated functions of all of the eukaryotic organelles are performed in the IM. Many of the membrane proteins that function in energy production, lipid biosynthesis, protein secretion, and transport are conserved in bacteria, but their cellular location is different. In bacteria, these proteins are located in the IM.

The IM is a phospholipid bilayer. In *E. coli* the principal phospholipids are phosphatidyl ethanolamine and phosphatidyl glycerol, but there are lesser amounts of phosphatidyl serine and cardiolipin. Other minor lipids include polyisoprenoid carriers (C55), which function in the translocation of activated sugar intermediates that are required for envelope biogenesis.

Trans-envelope Machines

Certain types of surface appendages such as flagella, which are required for bacteria motility; Type III secretion systems, which inject toxins into the cytoplasm of eukaryotic host cells during the infection process; and efflux pumps, which pump toxic molecules such as antibiotics from the cell clear across the cell envelope into the surrounding media and are responsible in part for much of the antibiotic resistance in pathogenic bacteria, are molecular machines that are made up of individual protein components that span the peptidoglycan and are located in all cellular compartments. The structures of some of these machines are known at sufficient resolution to provide meaningful insight into the size of the various cellular compartments in *E. coli*. these size predictions are in reasonable agreement. However, it should be noted that experimental measurements of the volume of the periplasm, for example, vary widely.

Envelope Biogenesis

All of the components of the Gram-negative cell envelope are synthesized either in the cytoplasm or at the inner surface of the IM. Accordingly, all of these components must be translocated from the cytoplasm or flipped across the IM. Periplasmic components must be released from the IM, peptidoglycan components must be released and polymerized, and OM components must be transported across the aqueous, viscous periplasm and assembled into an asymmetric lipid bilayer. All of this construction takes place outside of the cell in a potentially hostile environment that lacks an obvious energy source. It seems clear that there is no ATP out there for example.

All proteins, of course, are synthesized in the cytoplasm. Proteins destined for the periplasm or the OM are made initially in precursor form with a signal sequence at the amino terminus. The signal sequence targets them for translocation from the cytoplasm. This translocation reaction is catalyzed by an essential, heterotrimeric IM protein complex called SecYEG. The signal sequence and this heterotrimeric membrane protein complex are conserved throughout biology. The essential ATPase SecA, together with the proton motive force, drives this translocation reaction. Periplasmic and OM proteins are generally translocated in post-translational fashion, i.e., synthesis and translocation are not coupled. Proteins must be secreted in linear fashion from the amino to the carboxy terminus like spaghetti through a hole; SecYEG cannot handle folded molecules. The cytoplasmic SecB chaperone maintains these secreted proteins in unfolded form until they can be secreted. During the secretion process the signal sequence is proteolytically removed by Signal Peptidase I. Other components of the Sec translocon, such SecD,

SecF, and YajC, perform important but nonessential function(s) during translocation, perhaps facilitating release of secreted proteins into the periplasm. Once released, periplasmic proteins are home, but it seems likely that chaperones function to prevent misfolding and aggregation. For example, the periplasmic protein MalS, which contains disulfide bonds, requires the periplasmic disulfide oxidase DsbA for proper folding. In the absence of the DsbA the periplasmic protease/chaperone DegP (HtrA) can substitute.

Periplasmic chaperones function to protect OMPs during their transit through the periplasm. Three such proteins have been well characterized and shown have general chaperone activity: SurA, which also functions as a peptidyl-proline isomerase, Skp, and the aforementioned DegP. Genetic analysis indicates that these three proteins function in parallel pathways for OMP assembly; SurA functions in one pathway; DegP/Skp function in the other. Mutants lacking either one of these pathways are viable, but cells cannot tolerate loss of both. Mutants lacking SurA and Skp, or SurA and DegP are not viable and they show massive defects in OMP assembly. By definition then, these chaperone pathways are redundant. However, this redundancy does not reflect equal roles in OMP assembly. The major OMPs, which account for most of the protein mass of the OM, show preference for the SurA pathway as does the minor OMP LptD. At present no OMP that prefers the DegP/Skp pathway has been identified. It may be that many minor OMPs show no pathway preference. The primary role of the DegP/Skp pathway may be to rescue OMPs that have fallen off the normal assembly pathway, particularly under stressful conditions. It is also possible that other periplasmic proteins have chaperone function that is important for the assembly of a subset of OMPs.

The periplasmic chaperones deliver OMPs to a recently identified assembly site in the OM termed the Bam complex. This complex is composed of a large β-barrel protein, BamA (aka YaeT or Omp85), and four lipoproteins, BamBCDE (aka YfgL, NlpB, YfiO, and SmpA respectively). In addition to the β-barrel domain BamA has a large amino-terminal periplasmic domain composed of five POTRA (polypeptide transport associated). The structure of a large fraction of the BamA periplasmic domain has been determined. Each of the four visible POTRA domains has a nearly identical fold, despite the fact that the amino acid sequence identity between them is very low. In *E. coli* the first two POTRA domains are not essential for the life of the organism. Nevertheless, BamA is highly conserved in Gram-negative bacteria, and in these organisms there are always five POTRA domains. There are homologs of BamA in both mitochondria and chloroplasts, which are thought to be derived from Gram-negative bacteria. These homologs have one, two, or three POTRA domains, and the proteins function to assemble β-barrel proteins in the OM of these organelles. BamD is the only essential lipoprotein in the Bam complex, and it is highly conserved in Gram-negative bacteria as well. The remaining three lipoproteins are not essential, and they are conserved to varying degrees. We do not yet understand the mechanism of β-barrel folding nor do we understand the functions of the individual proteins in the Bam complex, but there is evidence suggesting that the POTRA domains of BamA may template folding by a process termed β augmentation.

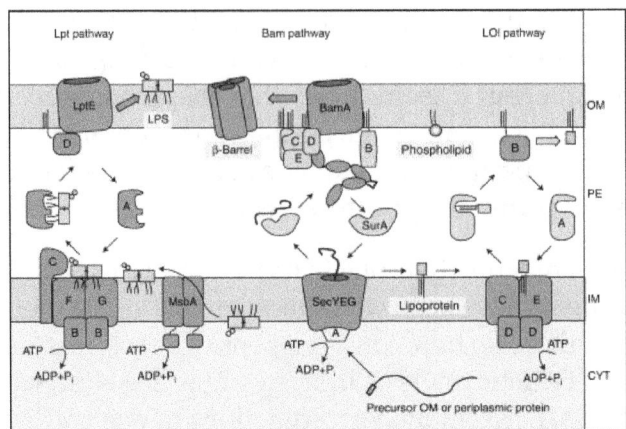

The cellular machineries required for OM biogenesis. The Lpt pathway, together with MsbA, transports LPS from its site of synthesis to the cell surface. β-barrel proteins and lipoproteins are made initially in the cytoplasm in precursor form with a signal sequence at the amino terminus. The signal sequence directs these precursors to the Sec machinery for translocation from the cytoplasm. Chaperones like SurA deliver beta-barrel proteins to the Bam machinery for assembly in the OM. For OM lipoproteins, after the signal sequence is removed and lipids are attached to the amino-terminal cysteine residue, the Lol machinery delivers them to the OM.

Lipoproteins are made initially with an amino-terminal signal sequence as well, and they too are translocated by the Sec machinery. However, the signal sequence is removed by a different signal peptidase, signal peptidase II. Signal sequence processing of lipoproteins requires the formation of a thioether diglyceride at the cysteine residue, which will become the amino terminus of the mature lipoprotein. Once the signal sequence is removed, an additional fatty acyl chain is added to the cysteine amino group. These lipid moieties tether the newly formed lipoprotein to the outer leaflet of the IM.

Some lipoproteins remain in the IM, and their biogenesis is complete after signal sequence processing and lipid addition. However, most of the lipoproteins in *E. coli* are destined for the outer membrane. The Lol system, which transports lipoproteins to the OM has been well characterized. There is an ABC transporter (LolCDE) in the IM that utilizes ATP hydrolysis to extract the molecule from the IM and pass it to a soluble periplasmic carrier called LolA. LolA delivers the molecule to the OM assembly site, which is the lipoprotein LolB. IM lipoproteins have a "Lol avoidance" signal so that they remain in the IM. The most common Lol avoidance signal is an aspartate residue at position two of the mature lipoprotein.

There is a second protein translocation system in the IM called Tat that translocates folded proteins. *E. coli* uses the Tat system for proteins which have prosthetic groups that must be added in the cytoplasm, and this constitutes a small fraction of the secreted proteins. Other bacteria, such as thermophiles, use the Tat system extensively;

presumable because it is easier to fold proteins in the cytoplasm than it is in the hostile environments they live in. In terms of components, the Tat system is remarkable simple; three components. TatB and TatC function to target proteins for translocation by TatA, but how this system recognizes that the substrate is folded, and how it accomplishes the translocation reaction are not yet understood.

Proteins destined for the IM are handled by the Sec machinery as well. However, in general, these proteins are targeted for cotranslational translocation by Signal recognition particle (SRP) and the SRP receptor FtsY. Presumably, posttranslational translocation of these hydrophobic substrates would be inefficient and perhaps dangerous, owing to their great potential for aggregation. Prokaryotic SRP is much simpler than its eukaryotic counterpart. It contains only a single protein, Ffh (fifty four homolog) and an RNA, Ffs (four point five S RNA). Both Ffh and Ffs are GTPases, as are their more complex eukaryotic counterparts.

The transmembrane α-helices, which are characteristic of most biological membrane proteins in both prokaryotes and eukaryotes, serve as secretion signals. The first trans-membrane segment functions as a signal sequence to initiate translocation of the sequences that follow it. These transmembrane sequences tend to be longer and more hydrophobic than typical signal sequences, and this serves as the basis for SRP recognition. This signal sequence is not cleaved; it remains attached serving now as a typical transmembrane helix. The second transmembrane helix functions to stop the translocation reaction and this helix exits the SecYEG translocator laterally where it remains in the IM. The third trans-membrane helix functions again as an uncleaved signal sequence. These alternating start and stop translocation signals stitch IM proteins into the membrane in stepwise fashion.

Small IM proteins, especially those with small periplasmic domains, can be inserted into the membrane by a second IM translocase called YidC. YidC family members can be found in mitochondria and chloroplasts. Like their mitochondrial homologs, YidC plays an important role in the assembly of energy-transducing membrane proteins such as subunit c of ATPase. YidC may also play a role in the SecYEG-dependent insertion of larger IM proteins during the lateral transfer of the trans-membrane α-helices into the lipid bilayer.

LPS, including the core polysaccharide, and the O-antigen are both synthesized on the inner leaflet of the IM. LPS is flipped to the outer leaflet of the IM by the ABC transporter MsbA. O-antigen is synthesized on a polyisoprenoid carrier, which then flips it to the outer leaflet. The O-antigen is ligated to the LPS core in the outer leaflet of the IM, a reaction catalyzed by WaaL. Note that the common laboratory strain *E. coli* K-12 does not make the O-antigen. Accordingly, it is termed "rough," as opposed to the wild-type "smooth" strain. In the last several years a combination of genetics, biochemistry, and bioinformatics was employed to identify seven essential proteins that are required to transport LPS from the outer leaflet of the OM to the cell surface. These proteins have been termed Lpt (lipopolysaccharide transport); LptA (aka YhbN), LptB (aka YhbG),

LptC (aka YrbK), LptD (aka Imp or OstA), LptE (RlpB), LptF (aka YjgP), and LptG (aka YjgQ). The large β-barrel protein, LptD and the lipoprotein LptE form a complex in the OM. LptA is made with a cleavable signal sequence and resides in the periplasm. LptF and LptG are IM proteins that likely interact with the cytoplasmic protein LptB, a predicted ATPase, to form an ABC transporter that together with the bitopic IM protein LptC, extracts LPS from the IM and passes it to the periplasmic protein LptA for delivery to the OM assembly site, LptD and LptE. An alternative model proposes that all seven proteins together form a transenvelope machine that transports LPS directly from the IM to the cell surface in analogy with efflux pumps. What is clear is that if any of the seven proteins are removed, LPS accumulates in the outer leaflet of the IM.

Like LPS, phospholipids are synthesized in the inner leaflet of the IM. MsbA can flip these molecules to the outer leaflet of the IM, but it is likely that other mechanisms to flip these molecules also exist. How phospholipids reach the OM is not known. What is known is that phospholipids added into the OM reach the IM very quickly. This is true even for lipids like cholesterol which are not naturally found in bacteria. This could suggest sites of IM-OM fusion, or hemi-fusion, that allow intermembrane phospholipid trafficking by diffusion, a hypothesis made by Manfred Bayer long ago that has remained highly controversial.

The Gram positive Cell Envelope

The Gram-positive cell envelope differs in several key ways from its Gram-negative counterpart. First and foremost, the outer membrane is absent. The outer membrane plays a major role in protecting Gram-negative organisms from the environment by excluding toxic molecules and providing an additional stabilizing layer around the cell. Because the outer membrane indirectly helps stabilize the inner membrane, the peptidoglycan mesh surrounding Gram-negative cells is relatively thin. Gram-positive bacteria often live in harsh environments just as *E. coli* does—in fact, some live in the gut along with *E. coli*—but they lack a protective outer membrane. To withstand the turgor pressure exerted on the plasma membrane, Gram-positive microorganisms are surrounded by layers of peptidoglycan many times thicker than is found in *E. coli*. Threading through these layers of peptidoglycan are long anionic polymers, called teichoic acids, which are composed largely of glycerol phosphate, glucosyl phosphate, or ribitol phosphate repeats. One class of these polymers, the wall teichoic acids, are covalently attached to peptidoglycan; another class, the lipoteichoic acids, are anchored to the head groups of membrane lipids. Collectively, these polymers can account for over 60% of the mass of the Gram-positive cell wall, making them major contributors to envelope structure and function. In addition to the TAs, the surfaces of Gram-positive microorganisms are decorated with a variety of proteins, some of which are analogous to proteins found in the periplasm of Gram-negative organisms. Because there is no outer membrane in Gram-positive organisms to contain extracellular proteins, all these proteins feature elements that retain them in or near the membrane. Some contain membrane-spanning helices and some are attached to lipid anchors inserted in the membrane. Others are covalently attached to or associated tightly with peptidoglycan. Still others bind to teichoic acids. Studies on *S. aureus* have shown that the composition of surface-expressed proteins can change dramatically depending on environmental cues or growth conditions, reflecting the important role of the cell envelope in adapting to the local environment.

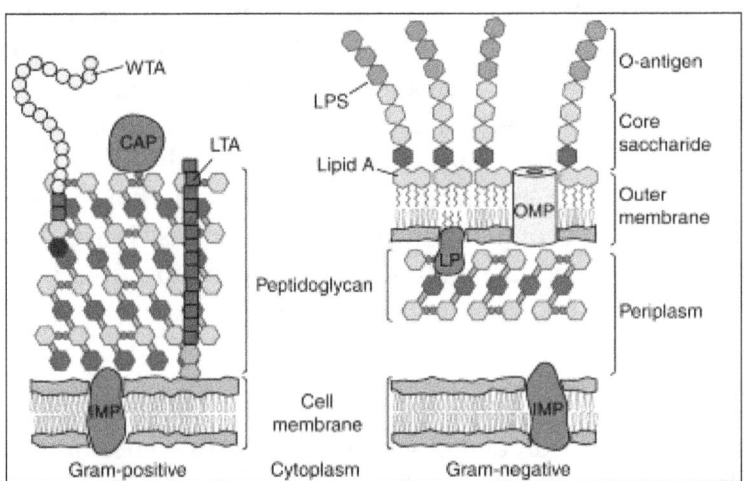

Depiction of Gram-positive and Gram-negative cell envelopes: CAP = covalently attached protein; IMP, integral membrane protein; LP, lipoprotein; LPS, lipopolysaccharide; LTA, lipoteichoic acid; OMP, outer membrane protein; WTA, wall teichoic acid.

Gram positive Peptidoglycan

The chemical structure of peptidoglycan in Gram-positive organisms is similar to that in Gram-negatives in that it is composed of a disaccharide-peptide repeat coupled through glycosidic bonds to form linear glycan strands, which are crosslinked into a meshlike framework through the peptide stems attached to the disaccharide repeat. The major difference between Gram-positive and Gram-negative peptidoglycan involves the thickness of the layers surrounding the plasma membrane. Whereas Gram-negative peptidoglycan is only a few nanometers thick, representing one to a few layers, Gram-positive peptidoglycan is 30–100 nm thick and contains many layers.

There are many differences among Gram-positive organisms with respect to the details of peptidoglycan structure, but perhaps the most notable difference relates to the peptide crosslinks between glycan strands. *S. aureus* contains crosslinks in which the peptides are connected through a pentaglycine branch extending from the third amino acid of one of the stem peptides. This pentaglycine branch is assembled by a set of nonribosomal peptidyl transferases known as FemA, B, and X. Staphylococci can tolerate, albeit with difficulty, the loss of FemA or B, but not of FemX, which attaches the first glycine unit to the stem peptide. Many Gram-positive organisms contain branched stem peptides, but *B. subtilis* does not; the stem peptides and crosslinks in this organism are identical in structure to those found in *E. coli*.

Branched stem peptides in *S. aureus* and other Gram-positive organisms play a variety of roles. Chief among these roles, they serve as attachment sites for covalently-associated proteins. They have also been implicated in resistance to beta lactam antibiotics. Beta lactams inactivate transpeptidases that catalyze the peptide crosslinking step of peptidoglycan synthesis by reacting with the active site nucleophile of transpeptidases. Transpeptidases that couple branched stem peptides are mechanistically similar to those that couple unbranched stem peptides; however, their substrate specificity is sufficiently different that they only recognize unbranched stem peptides, and some of them are resistant to beta lactams. For example, methicillin-resistant *S. aureus* strains express a transpeptidase, PBP2A, that couples *only*pentaglycine-branched substrates. Many other Gram-positive organisms are also thought to harbor low affinity PBPs that preferentially recognize and couple branched stem peptides. It is thus speculated that the evolution of branched peptides in the peptidoglycan biosynthetic pathway may be an adaptation that enables escape from beta lactams.

Surface Proteins

S. aureus colonizes human skin and mucosal surfaces. Breaches in the epithelium occasionally result in invasive *S. aureus* infections that are extremely serious. The ability to adhere to host tissue is a crucial first step in effective colonization by *S. aureus*, and a variety of surface factors are involved in this process. These factors include teichoic acids, as well as surface proteins that recognize components of host extracellular matrix such as fibronectin, fibrinogen, and elastin.

Some of these surface proteins, called adhesins, are attached via noncovalent ionic interactions to peptidoglycan or teichoic acids, but many are attached covalently to stem peptides within the peptidoglycan layers. Proteins destined for covalent surface display contain an amino-terminal signal sequence that enables secretion through the cytoplasmic membrane and a carboxy-terminal pentapeptide cell wall sorting motif, which is commonly LPXTG. Enzymes called sortases catalyze a transpeptidation reaction between these sorting motifs and the glycine branch of the stem peptide of peptidoglycan precursors. The transpeptidation reaction is thought to occur in two steps: in the first, a nucleophile in the sortase active site attacks the amide bond between the threonine and glycine of the sorting motif to produce a covalent intermediate in which the amino-terminal portion of the protein substrate is attached to the enzyme; in the second, the glycine branch of the stem peptide enters the active site and the nucleophilic amino terminus of the glycine branch attacks the acyl- enzyme intermediate, regenerating the enzyme and forming a new TG amide bond that anchors the protein to the peptidoglycan precursor. The protein-modified peptidoglycan precursor is then incorporated into peptidoglycan.

In *S. aureus*, more than twenty protein substrates for the major sortase, sortase A, have been identified. In addition to adhesins, these protein substrates include proteins involved in immune system evasion, internalization, and phage binding. A minor sortase,

sortase B, is responsible for surface display of proteins involved in iron acquisition, which is necessary for pathogenesis because iron is required for the function of many bacterial enzymes. In other Gram-positive organisms, sorting enzymes similar to SrtA and SrtB covalently couple proteins that comprise pili.

In addition to covalent attachment, Gram-positive organisms have other ways of retaining cell surface proteins. Many proteins involved in peptidoglycan biosynthesis are anchored to the cytoplasmic membrane by membrane spanning helices. Some proteins required for cell-wall degradation are associated noncovalently with peptidoglycan; others appear to be scaffolded and activated by teichoic acids or other types of cell surface polymers.

Teichoic Acids

Teichoic acids are anionic cell surface polymers found in a wide range of Gram-positive organisms, including *S. aureus* and *B. subtilis*. There are two major types of teichoic acids: wall teichoic acids (WTAs), which are coupled to peptidoglycan, and lipoteichoic acids (LTAs), which are anchored to the cell membrane. Wall teichoic acids are attached via a phosphodiester linkage to the C6 hydroxyl of occasional MurNAc residues in peptidoglycan. Although the structural variations are considerable, the most common WTAs are composed of a disaccharide linkage unit to which is appended a polyribitol phosphate (polyRboP) or polyglycerol phosphate (polyGroP) chain containing as many as 60 repeats. WTAs extend perpendicularly through the peptidoglycan mesh into what has been characterized as a "fluffy" layer beyond. *S. aureus* produces polyRboP WTAs; *B. subtilis* produces either polyRboP and polyGroP depending on the strain. The hydroxyls on the RboP or GroP repeats are tailored with other groups, typically d-alanyl esters or glycosyl moieties, and the nature and extent of the tailoring modifications significantly affect the properties and functions of WTAs.

LTAs are similar to WTAs in that they are composed of polyGroP polymers that are often functionalized with d-alanine or a sugar moiety; however, they also differ in a number of ways. For example, they contain glycerol-phosphate repeats of opposite chirality to those found in WTAs. Furthermore, rather than being attached to peptidoglycan, they are anchored to membrane-embedded glycolipids and typically contain fewer GroP repeats. Thus, they extend from the cell surface into the peptidoglycan layers rather than through and beyond. Together, the LTAs and WTAs comprise what has been concisely described as a "continuum of anionic charge" that originates at the Gram-positive cell surface but extends well beyond the peptidoglycan barrier. The importance of this continuum of negative charge is underscored by the fact that Gram-positive organisms lacking WTAs (either because they do not contain the gene clusters or because they are grown under phosphate-limiting conditions) produce other types of polyanionic polymers in which the negative charges are supplied by carboxylate or sulfate groups. Furthermore, although neither LTAs nor WTAs are essential, deleting the pathways for the biosynthesis of either of these polymers produces organisms that have cell division and morphological defects as well as other, less serious growth defects. Moreover, it is not possible to delete both pathways because the genes involved are synthetic lethals.

Teichoic acids account for a significant fraction of the cell wall mass in producing organisms and their functions are many, varied, and species-dependent. Because they are anionic they bind cations and thus play a role in cation homeostasis. Networks of metal cations between WTAs also influence the rigidity and porosity of the cell wall. The negative charge density on WTAs can be modulated by tailoring modifications that introduce positive charges along the polymer backbone, and these modifications can have a profound effect on the interactions of bacteria with other cells or molecules. For example, in *S. aureus*, a d-alanine transferase couples d-alanine moieties to free hydroxyls on the polyribitol phosphate backbone. *S. aureus* strains lacking d-alanine esters are more susceptible to antimicrobial cationic peptides and to lytic enzymes produced by host neutrophils. They also display reduced autolysin activity, suggesting a role for functionalized WTAs in scaffolding or activating hydrolytic enzymes involved in cell wall synthesis and degradation.

Cell Envelope of Corynebacterineae

The *Corynebacterineae* are a group of bacteria that includes the very important pathogens *Mycobacterium tuberculosis* and *Mycobacterium leprae*. These bacteria are generally classified as high G+C Gram positives, however their cell envelope has characteristics of both Gram-positive and Gram-negative bacteria. Indeed, a genome-based phylogeny places them in between Gram positives and Gram negatives.

The cell envelope of the *Corynebacterineae* is very complex and this complexity contributes substantially to their virulence. The peptidoglycan layer that surrounds a standard IM contains covalently attached arabinogalactan and this is covalently attached to mycolic acids. These mycolic acids have very long alkyl side chains (up to C_{90}) that give the bacteria a waxy appearance and account for their resistance to acid decolorization during staining procedures (acid-fast).

Unlike other Gram-positive bacteria, *Corynebacterineae* have an OM. This OM appears to be symmetrical unlike the Gram-negative OM. Mycolic acids appear essential for this OM, but how they are organized remains unclear. Mycobacteria have porin proteins in their OM, but the structure of the main porin, MspA, from *Mycobacterium smegmatis* is quite different from the typical porin proteins of Gram-negative bacteria. Indeed, mycobacteria have no obvious homologs of the Bam complex members. This might suggest a novel mechanism for the assembly of proteins in the OM of *Corynebacterineae*.

Permeability Barrier of Gram negative Cell Envelopes

Gram-negative bacteria are intrinsically resistant to many antibiotics. Species that have acquired multidrug resistance and cause infections that are effectively untreatable present a serious threat to public health. The problem is broadly recognized and tackled at both the fundamental and applied levels.

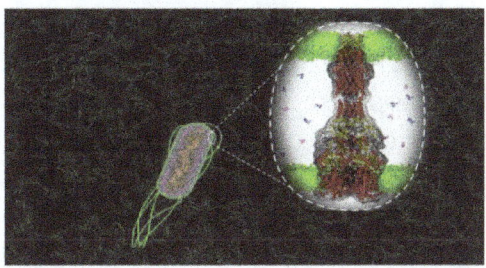

Gram-negative Pathogens and Challenges of Antibiotic Discovery

Drug resistance presents an ever-increasing threat to public health and encompasses all major microbial pathogens and antimicrobial drugs. Some pathogens have acquired resistance to multiple antibiotics and cause infections that are effectively untreatable. Among pathogenic Gram-negative *Enterobacteriaceae*, *Acinetobacter*, and *Pseudomonas*, species have emerged that are resistant to all good antibiotics. One of the most troubling event is the worldwide spread of carbapenem-resistant *Klebsiella* spp.Infections caused by these resistant variants have a mortality rate of up to 50%. By 2013, 17% of *Escherichia coli*infections became multidrug resistant. In some regions, fluoroquinolones are no longer on the lists of recommended treatment options. Such environmental species as *Pseudomonas aeruginosa*, *Stenotrophomonas malto-philia*, *Burkholderia cepacia* complex (Bcc), and *Acinetobacter baumannii* are intrinsically resistant to antibiotics. Among these species, *P. aeruginosa* is a common nosocomial pathogen, the causative agent of many life-threatening infections and the major reason for the shortened life span of people with cystic fibrosis (CF). *P. aeruginosa* infections can be successfully treated by only a few specific representatives of fluoroquinolones, β-lactams, or aminoglycosides. However, even these few antibiotics fail against antibiotic-resistant *P. aeruginosa* isolates. Thus, there is a strong need for new therapeutic options, particularly those directed against multiresistant Gram-negative bacteria.

The discovery of new antibiotics effective against Gram-negative bacteria is a major challenge, primarily because of a low hit rate during screening of compound libraries, which is up to 1000-fold lower in *P. aeruginosa* than against Gram-positive bacteria. The major reasons for such a low hit rate are the low permeability barrier of two-membrane cell envelopes of Gram-negative bacteria and insufficient chemical diversity of compound libraries to probe this barrier. Gram-negative bacteria vary significantly in their permeability to antibiotics, but one could expect that the basic principles established by extensive studies of *E. coli* would apply equally to such "impermeable" species as *Burkholderia* spp. or *Pseudomonas* spp. It remains unclear, however, whether permeation rules, in analogy with Lipinski's rules, if such existed and were applied to structure–activity relationships or to filtering compound libraries, would yield compounds that permeate all Gram-negative barriers.

The Two-membrane Barrier of Gram negative Bacteria

The susceptibility of Gram-negative bacteria to antibiotics is defined by two opposing fluxes across the two membranes of these species. The influx and uptake of antibiotics are significantly slowed by the elaborate outer membrane (OM). This membrane is an asymmetric bilayer of lipopolysaccharides (LPS) and phospholipids, into which nonspecific porins and specific uptake channels are embedded.The LPS-containing bilayers are more rigid than normal bilayers, slowing passive diffusion of hydrophobic compounds, whereas narrow pores limit by size the penetration of hydrophilic drugs. The slow influx of drugs across the OM is further opposed by active efflux mediated by multidrug efflux transporters. Multidrug efflux transporters are structurally and functionally diverse, with some transporters pumping antibiotics across the inner membrane and reducing concentrations of antibiotics in the cytoplasm, whereas others expel antibiotics from the periplasm into the external medium. The latter transporters confer resistance to antibiotics by associating with the periplasmic and OM accessory proteins to form trans-envelope complexes. These complexes enable conversion of the energy stored in the inner membrane into active efflux of antibiotics across the OM. Efflux of antibiotics across the inner membrane acts synergistically with the trans-envelope efflux and, as a result, inactivation of efflux pumps leads to dramatic sensitization of Gram-negative bacteria to antibiotics. The clinical relevance of efflux of multiple antibiotics has also been established. For example, in clinical isolates of *E. coli* and *Klebsiella pneumoniae*, fluoroquinolone resistance is linked to overproduction of the AcrAB-TolC efflux pump, whereas multiple efflux pumps confer antibiotic resistance in *P. aeruginosa*, *Burkholderia* spp., and *A. baumannii*. The interplay between uptake and efflux defines the steady-state accumulation level of antibiotics at targets.

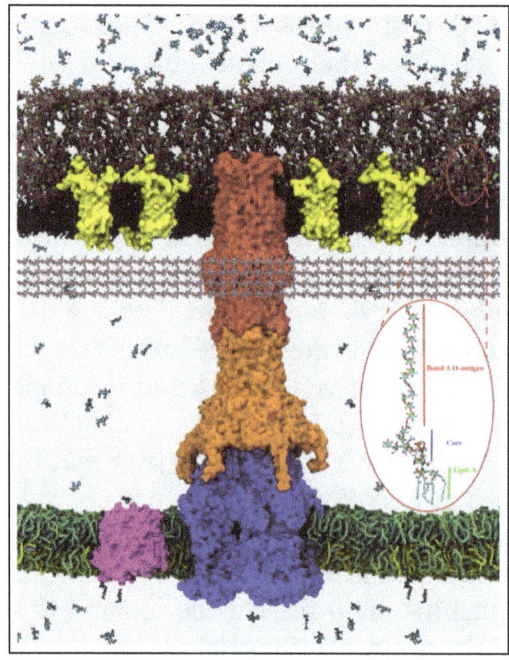

The outer leaflet of the outer membrane is assembled of LPS (pink color) corresponding to the band A antigen,25 and the inner leaflet contains glycerophospholipids 1,2-dipalmitoyl-sn-glycero-3-phosphoe-thanolamine (DPPE). Green spheres are magnesium ions bound to O-chains and core polysaccharides of LPS. The inner membrane contains an equimolar mixture of cardiolipin, 1-palmitoyl-2-oleoyl-sn-glycero-3-phosphoethanolamine (POPE)b and 1-palmitoyl-2-oleoyl-sn-glycero-3-phosphoglycerol (POPG). Embedded in the outer membrane are the main porin OprF (PDB 4RLC) in a yellow surface density. In the inner bilayer is localized MdfA transporter, represented in a magenta surface (PDB 4ZOW). The modeled structure of the assembled MexAB-OprM multidrug efflux pump (MexB, blue; MexA, orange; OprM, red) spans both the inner and outer membranes. The structure of MexAB-OprM is from a long (1 μs) all-atom molecular dynamics simulations with the tripartite complex embedded in 1-palmitoyl-2-oleoyl-sn-glycero-3-phosphocholine (POPC) bilayers mimicking inner and outer membranes. The MexA component lacks the N-terminal lipid modification. The outer and inner membranes are based on separate protein-free all-atom MD simulations with the above composition. Other structures were taken from the Protein Data Bank. Ciprofloxacin molecules were added to illustrate a difference of concentrations created by slow diffusion across the outer membrane through porins and LPS-containing bilayer and the active efflux across the outer (MexAB-OprM) and inner (MdfA) membranes. (Inset) Single representative LPS molecule from the simulations. The peptidoglycan is pictorially represented by fiber-like structures underneath the outer membrane.

Composition and Properties of LPS-containing Bilayers

Despite a similar structural organization, outer membranes of Gram-negative bacteria differ dramatically in their permeability properties. These differences are largely attributed to differences in the permeability properties of general porins, but the chemical structure and properties of the LPS-containing bilayer also play an important role. Typical LPS comprises a basic lipid A structure containing an N- and O-acylated diglucosamine bisphosphate backbone. Chemical variations of lipid A among species involve the number of primary acyl groups and the types of fatty acids substituting the primary and secondary acyl groups. *Escherichia* lipid A is most frequently described as a hexaacylated molecular species, although penta-and tetra-acylated molecules are also present in various amounts. Most of the laboratory-adapted strains of *P. aeruginosa* synthesize a penta-acylated (band A, 75% of the molecules) LPS, with some proportion made as a hexa-acylated LPS (25% of the molecules). Growth conditions, notably magnesium levels, can affect the acylation pattern of *P. aeruginosa* lipid A. Among isolates from chronically infected CF patients, which are known to be mutants generally unable to synthesize O-antigen side chains, a hexa-acylated LPS form predominates, although a hepta-acylated lipid A has been isolated, containing an additional palmitoyl (C16:0) group linked to the primary 3-hydroxyde-canoic acid group at position 3′ of glucosamine 2. The hexa- and hepta-acylated lipid A moieties also contain cationic 4-amino-4-deoxy-L-arabinose sugars. Similarly, the major lipid A species in *Burkholderia* spp.

(*B. cepacia, B. mallei, B. pseudomallei*) consists of a biphosphorylated disaccharide backbone, but it is constitutively modified with 4-amino-4-deoxy-L-arabinose and penta- or tetra-acylated. Lipid A of *A. baumannii* is often modified by phosphorylethanolamine and an unusual sugar galactosamine and hexa- and hepta-acylated.

		P. aeruginosa	A. baumannii	B. cepacia ATCC 17759, ATCC 25416
O-antigen		**Lab strains**: EPS, S-type LPS **CF strains**: R-type	**Lab and clinical strains**: EPS; R-type dominates, loss of LPS in resistant strains	**Lab and CF strains**: both S- and R-types present
Outer core		Alanine	Alanine	Phosphoethanolamine
Inner core		Carbamoyl phosphate on HepII; a tri-phosphate on HepI;	Ko in some strains; Kdo tetrasaccharide; short rhamnoglycans; No phosphates	Ko replaces KdoII; 4-deoxy-4-amino-arabinose attached to Ko or replaces Ko non-stoichimetrically
Lipid A		**Lab strains**: C12/C10 chains, penta-acylated; hexa-acylated with C16, 4-deoxy-4-amino-arabinose linked to phosphates at low Mg²⁺; **CF strains**: hexa-acylated with C16; hepta-acylated	**All strains**: C12/14 fatty acid chains; hepta-acylated; **Clinical isolates**: phosphoethanolamine, galactosamine	**All strains**: C14/C16 fatty acid chains, tetra- and penta-acylated; 4-deoxy-4-amino-arabinose linked to phosphates

Diversity of chemical structures and modifications in LPS.

The P. aeruginosa band A LPS molecule is shown, in the above figure, for comparison. Abbreviations: Kdo, α-3-deoxy-D-manno-oct-2-ulosonic acid; Hep, heptulose; NGal, galactosamine; EPS, extracellular polysaccharide; S-type, "smooth" LPS containing O-chains; R-type, "rough" LPS lacking O-chains; Ko, D-glycero-D-talo-oct-2-ulosonic acid.

Early studies showed that the permeabilities of *E. coli* and *P. aeruginosa* OM to hydrophobic steroid probes are similar, suggesting that the differences in the lipid A acylation state and the length of fatty acids do not affect significantly the OM permeability to small planar molecules. On the other hand, amphiphilic and charged molecules are likely to interact with the backbone of lipid A and LPS cores, and their permeation could be sensitive to modifications of both the lipids and polysaccharides of the LPS-containing bilayers.

LPS cores of enterobacteria typically consist of 8–12 often branched sugar units. The sugar at the reducing end is always α-3-deoxy-D-manno-oct-2-ulosonic acid (Kdo) (2→6)-linked to lipid A. At the C-4 position of Kdo, there may be one or two Kdo groupings. Three L-glycero-D-manno-heptose residues are also (1→5)-linked to the first Kdo. A heptose residue may be substituted by a phosphate, pyrophosphate, or phosphorylethanolamine group or by another sugar to make up the inner core. In *Pseudomonas* spp. cores, one often finds an alanyl group substituting a galactosamine residue and a carbamoyl group on heptose I. *P. aeruginosa* PAO1 serotype O5 and its two rough-type mutants share these characteristics but have, in addition, three phosphomonoester groups on their heptose. It was suggested that these groups play a role in interactions with antibiotics because their presence correlates with resistance to β-lactams but not to aminoglycosides. Some *Acinetobacter* and *Burkholderia* spp. produce LPS cores devoid of heptose. *B. cepacia* and *A. hemolyticus* synthesize and incorporate the Kdo analogue D-glycero-D-talo-oct-2-ulosonic acid (Ko).

The O-chains determine the specificity of each bacterial serotype. A combination of monosaccharide diversity, the numerous possibilities of glycosidic linkage, substitution, and configuration of sugars, and the genetic capacities of the diverse organisms have all contributed to the uniqueness of the great majority of O-chain structures.

It is broadly accepted that LPS cores and lipid A of enterobacteria and *P. aeruginosa* are modified in response to specific growth conditions and stresses. These modifications are critical under the low magnesium ion conditions that destabilize the LPS leaflet and increase permeability of the OM. They also play an important role in the development of resistance against cationic antimicrobial peptides and polymyxins. However, in *Acinetobacter* spp. and *Burkholderia* spp. some of the modifications that reduce the negative-charge character of LPS are constitutive. How the phosphate, pyrophosphate, or phosphorylethanolamine groups on Kdo and heptoses, terminal sugars, amino acid, and other groups occurring nonstoichiometrically correlate with specific growth conditions of these bacteria remains unclear. Even less is known of how these modifications affect the packing and rigidity of the LPS layer and its interactions with proteins and small molecules in the context of the outer membrane.

Recent advances in molecular dynamics (MD) simulations of asymmetric bilayers containing LPS offer first glimpses into the packing and dynamics of such structures.The inner leaflet of the outer membrane is composed of phospholipids with a composition similar to that of the cytoplasmic membrane. Hence, the geometry of the LPS leaflet should match that of the phospholipid leaflet. On the basis of MD simulations, LPS-containing bilayers appear to be more disordered and thinner than bilayers assembled from phospholipids only. Magnesium ions form a layer of ionic bonds with phosphoryl moieties of lipid A and the core and stabilize LPS molecules in the bilayers. Also, LPS changes the hydration profile at membrane–water interfaces.The packing and rigidity of LPS bilayers are expected to affect permeation of amphiphilic and hydrophobic antibiotics. However, this picture remains incomplete because about 50% of the *E. coli* OM mass consists of protein and the OM resembles a LPS–protein aggregate. Changes in hydration and electrostatic profiles due to LPS may affect the environment of embedded proteins and influence protein–protein interactions.

Outer Membrane Proteins and Permeability

In *E. coli*, a few integral membrane proteins, such as OmpA and the general porins OmpF/C, are expressed at high levels. Besides these, there are minor proteins whose synthesis in some cases is strongly induced when they are needed, such as specific porins (e.g., PhoE and LamB), TonB-dependent receptors (e.g., FhuA and FepA), components of several protein export systems, proteins involved in the biogenesis of flagella and pili, and enzymes (e.g., OmpT protease and phospholipase A).The permeability properties of *E. coli* membranes are largely defined by general porins that have an exclusion limit at about 600 Da for OmpF. Most hydrophilic and amphiphilic antibiotics reach the periplasm through porins, whereas larger antibiotics, for example, erythromycin or novobiocin, are believed to cross the outer membrane through lipid bilayer by diffusion, which is expected to be slow.

Table: Major Porins and Efflux Pumps of Representative Gram-Negative Bacteria

Species	Relative OM permeability (%)	Major general porins	Major efflux pumps
E. coli	100	OmpF/OmpC	AcrAB-TolC
P. aeruginosa	1–8	OprF	MexAB-OprM, MexXY-OprM
B. cepacia	11	OpcP1/OpcP2	AmrAB-OprA, BpeAB-OprB, BpeEF-OprC
A. baumannii	1–5	OmpA-AB	AdeABC and AdeIJK

The existing paradigm that only small molecules (MW ~ 600 Da) can cross the membrane by passive diffusion has recently been challenged by studies of the CymA channel from *Klebsiella oxytoca*.CymA mediates diffusion of cyclodextrins with diameters up to 15 Å by a novel mechanism, which involves a mobile N-terminal peptide acting as a periplasmic gate. Binding of the incoming substrate displaces the N-terminal constriction and allows diffusion of the bulky molecule without compromising the permeability barrier of the OM. One could imagine that antibiotics could be modified in a way to mimic the interactions of substrates with specific porins and facilitate their own diffusion across the OM.

In addition, some leakage of large antibiotics is possible through export systems. *E. coli* cells lacking the TolC channel, which is involved in efflux of antibiotics and export of proteins as a part of the type I secretion pathway, are more resistant to vancomycin (2–4-fold increase in minimal inhibitory concentrations) than the wild-type cells, suggesting some penetration of this ~1200 Da molecule through TolC.Interestingly, this penetration requires the presence of active AcrAB transporter, further suggesting that vancomycin slips through TolC when it is engaged by the AcrAB pump. It is possible that permeation through other specific porins and TonB-dependent channels and the Bam complex responsible for the assembly of the outer membrane and at protein–LPS interfaces also contribute to the intracellular accumulation of antibiotics.

The overall compositions and architectures of *Pseudomonas* spp. and *Acinetobacter* spp. outer membranes are similar to those of other Gram-negative bacteria. About 160 various proteins are embedded into the asymmetric bilayer and play a role in generalized and specific uptake and export of various compounds and polypeptides. Unlike enterobacteria, outer membranes of Pseudomonads and *Acinetobacter* spp. do not contain general trimeric porins such as OmpF and OmpC of *E. coli*. It is estimated that the permeability of *P. aeruginosa* and *A. baumannii* outer membranes is only 1–8% of that *E. coli* and that nonspecific "slow" porins OprF and OmpA-AB, respectively, restrict access of molecules larger than ~200 Da, which is the size of a typical monosaccharide. However, the growth rate of *P. aeruginosa* as well as of *A. baumannii* in a rich medium is comparable to that of *E. coli*, and hence to sustain the high rate of metabolism, the net influx of nutrients is expected to be comparable for the three species. *P. aeruginosa*likely solves this conundrum of the low permeability of the outer membrane despite a high nutrient uptake in several ways: (i) by secreting degrading enzymes that convert larger molecules into monounits (pseudomonads are masters of biodegradation) and (ii) by producing a larger fraction of substrate-specific channels (such as OprB and OprD).

OprF is one of the most abundant proteins in *P. aeruginosa* with a copy number of 200 000 per cell. This protein plays a structural role by associating with LPS and peptidoglycan. *P. aeruginosa* mutants deficient in OprF synthesis have an almost spherical appearance, are shorter than wild-type cells, and do not grow in low-osmolarity medium. OprF also plays an important nonspecific uptake function by existing in closed and open conformations (400 open conformers of 200 000 total OprF).Much less is known about the major porin OmpA of *A. baumannii*. This protein has a permeability similar to that OprF and also plays both structural and uptake roles in the outer membrane, but whether or not it exists as different conformers is unclear.

In addition to "slow" porins, *P. aeruginosa* and *A. baumannii* use specific porins for uptake of small molecules. The large number of specific porins provides an advantage in nutrient-deficient environments, but could be limiting in rich growth media.*P. aeruginosa* OprB, specific for glucose uptake, and OprD, specific for the diffusion of basic amino acids and peptides, are best characterized. OprD is the primary channel for the entry of carbapenems across the OM, and the reduced expression or loss of OprD has been frequently observed in carbapenem-resistant clinical isolates. *A. baumannii* produces an OprD homologue CarO, but whether or not this porin provides a path for carpenems remains under debate.

Interestingly, although the outer membrane permeability of *A. baumannii* to cephalothin and cephaloridine, measured in intact cells, was found to be about 100-fold lower than that of *E. coli* K-12,these cells are more susceptible to large antibiotics such as novobiocin and erythromycin. Many *Acinetobacter* spp. are capable of using long-chain hydrocarbons as growth substrates, suggesting that their LPS–phospholipid bilayers of the outer membrane are more permeable for the hydrophobic molecules.

Outer membranes of *Burkholderia* spp. contain a trimeric general porin Omp38 (OpcP), a homologue of *E. coli* OmpF. The two studies of Omp38 permeability produced contradictory results: this porin is either similar to *E. coli* OmpF or by 1 or 2 orders of magnitude less permeable than OmpF. Further studies are needed to analyze the properties of porins and outer membranes for challenging Gram-negative bacteria such as *A. baumannii* and *Burkholderia* spp.

Efflux Pumps

All Gram-negative bacteria examined so far contain at least one multidrug efflux transporter responsible for protection against a variety of antimicrobial agents. AcrAB-TolC of *E. coli* is the best characterized efflux pump, and its homologues are broadly represented in enterobacteria and other Gram-negative species.In this three-component complex, AcrB is a proton-motive force driven transporter from the resistance–nodulation–division (RND) superfamily of proteins, TolC is an outer membrane channel, and AcrA is a periplasmic membrane fusion protein (MFP). The three proteins form a trans-envelope complex that expels multiple antibiotics from *E. coli* cells.Extensive structural and functional analyses, including our own studies, showed that AcrB captures its substrates from the membrane and from the periplasm and expels them through the TolC channel into the external medium with the help of AcrA protein. At least two substrate binding sites in AcrB, the proximal and the distal, are both located in the periplasmic domains of the protein. Thus, unlike other typical secondary transporters, AcrB with the help of AcrA and TolC expels substrates across the outer membrane while driven by the proton transfer across the cytoplasmic membrane. All RND-type multidrug efflux pumps are likely to share this mechanism, but it remains unclear whether some of these protein complexes can actually pump substrates not only across the outer but also across the inner membrane, thus performing the trans-envelope efflux of antibiotics from the cytoplasm directly into the external medium.

Model of the assembled MexB-MexA-OprM multidrug efflux pump from P. aeruginosa.

MexA and other MFPs are proposed to translate the conformational changes in MexB transporter driven by a proton-motive force into opening of the outer membrane channel OprM that enables efflux across the outer membrane. The tripartite complex shown is a snapshot of a 1 μs MD simulation of the complex embedded in a bilayer composed of 1-palmitoyl-2-oleoyl-sn-glycero-3-phosphocholine (POPC). Proteins are colored according to secondary structure. For visual clarity water molecules are not shown. Accompanying drugs are shown for illustration purposes only (blue, rifampicin; pink, erythromycin). The gray shape corresponds to the AcrAB-TolC density obtained by the cryo-EM imaging. The equilibrated MexAB-OprM complex in the bacterial envelope maintains the expected distance of the P. aeruginosa periplasm. Image was initially exported from VMD Molecular Viewer in STL format and posteriorly rendered in Blender. The gray shape corresponds to the AcrAB-TolC density obtained by the cryo-EM imaging.

Among AcrAB-TolC substrates are organic solvents, antibiotics, detergents, dyes, and even hormones. Examination of chemical structures, as well as cocrystallization and molecular dynamics studies, significantly advanced the understanding of the molecular mechanism of multidrug recognition by AcrB. However, physicochemical properties of compounds that would distinguish AcrB substrates from nonsubstrates remain elusive. The current consensus is that AcrAB-TolC is the most effective against amphiphilic compounds that diffuse slowly across the OM. Recent attempts to determine the affinity of AcrAB-TolC toward its substrates demonstrated convincingly that minimal inhibitory concentrations of antibiotics is a very poor, if at all, measure of whether an antibiotic is a good or a bad substrate of an efflux pump. Several uptake assays were developed to analyze drug efflux, with most of them informative in both *E. coli* and *P. aeruginosa*. However, further characterization of the kinetic behavior of efflux pumps in the context of two-membrane envelopes of Gram-negative bacteria is needed.

P. aeruginosa MexAB-OprM is a close homologue of AcrAB-TolC and the major "housekeeping" efflux pump, but it is not the only pump that is expressed and contributes to antibiotic resistance in this species. The constitutive expression of MexAB-OprM confers intrinsic antibiotic resistance, whereas elevated expression of MexXY-OprM is the major cause of aminoglycoside resistance in the absence of modifying enzymes. In addition, efflux pumps MexCD-OprJ and MexEF-OprN are often overproduced in multiresistant clinical isolates. MexEF-OprN and MexGHI-OprD play important roles in *P. aeruginosa* physiology and are involved in the secretion of quorum-sensing signals and biofilm formation. Furthermore, *P. aeruginosa* lacking either MexAB or MexHI is attenuated in animal models.

As in the case of *P. aeruginosa*, genomes of *Burkholderia* spp. and *Acinetobacter* spp. contain multiple operons encoding RND-type efflux pumps. Some of these pumps are constitutively expressed and confer intrinsic antibiotic resistance, whereas others are inducible and expressed under specific physiological conditions. AmrAB-OprA *B. pseudomallei* and its homologues in other *Burkholderia* spp. are largely responsible for intrinsic antibiotic resistance. AmrAB-OprA is closely related to MexXY-OprM of *P. aeruginosa* and, in addition to various antibiotics, confers resistance to aminoglycosides. In the most comprehensive genetic analysis, deletion of the 16 putative RND operons from *B. cenocepacia* strain J2315 showed that these pumps play differential roles in the drug resistance of sessile (biofilm) and planktonic cells. These studies revealed that (1) RND-3 (a homologue of AmrAB-OprA) and RND-4 play important roles in resistance to various antibiotics, including fluoroquinolones and aminoglycosides, in planktonic populations; (2) RND-3, RND-8, and RND-9 protect against the antimicrobial effects of tobramycin in biofilm cells; and (3) RND-8 and RND-9 do not play a role in ciprofloxacin resistance.

The *A. baumannii* genome encodes seven RND-type efflux pumps. AdeIJK confers intrinsic resistance to various antibiotics including β-lactams, fluoroquinolones, tetracyclines, lincosamides, and chloramphenicol. In addition, AdeABC and AdeFGH are overproduced in clinical multidrug resistant isolates. Surprisingly, overproduction of AdeABC and AdeIJK could also alter bacterial membrane composition, resulting in decreased biofilm formation but not motility. When expressed in *E. coli* at levels comparable to those of AcrAB-TolC, AdeIJK but not AdeABC was more effective in the removal of lipophilic β-lactams, novobiocin, and ethidium bromide.

Efflux across the Cytoplasmic Membrane

Highly abundant in genomes of various Gram-negative bacteria, single-component transporters extruding drugs from the cytoplasm into periplasm remain understudied in clinical settings, in part because their drug efflux activities are important only in the context of intact OM and in the presence of active efflux across the OM. Inactivation of single-component transporters that pump antibiotics across the inner membrane usually does not lead to significant changes in susceptibilities to antibiotics. However, inactivation of multiple such pumps in *E. coli* is able to negate even the activity of AcrAB-TolC.

Drug-specific transporters, such as TetA, are also spread by plasmids, and their contribution could be either additive or multiplicative depending on the combination of transporters expressed in host cells. Experimental data and kinetic modeling showed that the coexpression of one- and three-component transporters leads to a synergistic loss of susceptibility to shared substrates.

One-component transporters can belong to one of the three protein families: small multidrug resistance (SMR), major facilitator superfamily (MFS), or multidrug and toxic extrusion (MATE). *E. coli* SMR transporter EmrE, MFS MdfA, and MATE transporter NorE are the best characterized representatives. MdfA and NorE are commonly expressed in fluoroquinolone-resistant isolates and contribute to resistance against these antibiotics. The properties and mechanisms of these proteins are well-characterized, but very little is known about their homologues in "difficult" Gram-negative pathogens.

Approaches to Bypass or Break the Permeability Barrier

In general, several approaches are envisioned as to how antibiotic penetration across the permeability barrier of Gram-negative bacteria could be improved, but all of these approaches suffer from intrinsic limitations that need to be further addressed for new therapeutics to emerge.

Inhibition of New Accessible Targets

The most traditional approach would be to identify new accessible targets on cell surfaces or in the periplasm that could be interrogated in drug discovery efforts. The major bottleneck here is that such targets are optimized over millions of years of evolution to be resistant to antibiotics. On the other hand, recent breakthroughs in understanding the molecular mechanisms of the cell envelope assembly promise that such a goal is attainable, albeit through a more target-oriented approach. LPS and its biosynthesis pathway have been considered highly attractive targets in Gram-negative bacteria for decades. Polymyxins, cationic cyclic lipopeptides, bind to LPS and permeabilize OM. These peptides recently re-emerged in clinics to treat multidrug-resistant Gram-negative infections, and their less toxic and "improved" variants attract significant attention. *Burkholderia* spp. are intrinsically resistant to cationic peptides, but they have high efficacy against susceptible enterobacteria, *P. aeruginosa* and *A. baumannii*. There are two drug candidates in clinical trials that target the LPS pathway in *P. aeruginosa*. ACHN-975 (Achaogen) is a synthetic molecule that inhibits LpxC deacylase, and POL7080 (Polyphor) is a synthetic peptide targeting the outer membrane protein LptD, involved in exporting LPS molecules across the periplasm. In addition to targeting essential enzymes, these inhibitors are expected to be synergistic with other antibiotics by enabling their permeation across the cell envelope.

Identification of Uptake Pathways and the "Trojan Horse" Approach

Another approach to bypass the permeability problem is to achieve fast or facilitated uptake of an antibiotic. Such efficient uptake would negate the impact of efflux and increase the concentration of an antibiotic at the target. *P. aeruginosa* and other "impermeable" Gram-negative species rely on specific porins and an active uptake system to fulfill their metabolic demands and to maximize their growth rates. These pathways could be potentially exploited for delivery of antibiotics to their targets through the characterization of specificity determinants and modification of antibiotics to enable their uptake through such specific systems. In the recent study of carbapenem penetration into *P. aeruginosa*, molecular metadynamics simulations were used to delineate the lowest energy path of native substrates and carbapenem antibiotics through OccD1(OprD) and OccD3 channels and to identify molecular features in antibiotics that enable diffusion through specific channels. Subsequent synthesis of carbapenem analogues to chemically probe some of the required features of permeation while maintaining the inherent activity led to an analogue of meropenem, for which a simple substitution on the side chain resulted in diminished dependence on OccD1 for translocation. This study is the first success story and demonstrates that a rational design of compounds with enhanced OM permeability is feasible. The major limitation of such an approach is a high frequency of resistance because the pathways are redundant and nonessential. However, identification and exploitation of uptake systems that are essential during establishment and proliferation of infections could facilitate the development of new therapeutics.

The "Trojan Horse" approach, which is largely derived from siderophore-conjugated antibiotics, is an example of such a strategy. Siderophore-conjugated antibiotics can be actively transported into the periplasm and cytoplasm by various iron-uptake systems. A common pathway of bacterial siderophore transport systems in Gram-negative bacteria has been identified. An outer membrane transporter binds the Fe^{3+}–siderophore complex with an affinity in the range of 1 nM and translocates this complex across the outer membrane with the help of TonB protein anchored in the cytoplasmic membrane. This process is driven by the cytoplasmic membrane potential. The Fe^{3+}–siderophore in the periplasm is bound by a binding protein that delivers its cargo via the cognate ABC transporter into the cytoplasm. Alternatively, the reduced iron is unloaded in the periplasm, as in the case of yersiniabactin and *P. aeruginosa*pyoverdines. Hence, siderophore-conjugated antibiotics could be delivered by such uptake pathways either into the periplasm or all the way into the cytoplasm. BAL30072 (Basilea) is the only candidate currently in clinical trials. This compound is a monocyclic β-lactam conjugated to a dihydropyridone siderophore moiety, which has a potent activity against multidrug resistant Gram-negative pathogens including "impermeable" species.

Studies of natural peptide antibiotics suggested additional pathways that could be exploited for bypassing the "impermeable" cell envelopes. Pacidamycins are uridyl peptide antibiotics, which specifically kill strains against *P. aeruginosa*. These antibiotics inhibit MraY translocase by catalyzing the first step in peptidoglycan synthesis. With MW > 800 Da, pacidamycins cannot penetrate the OM but exploit a specific uptake system to reach the target. The ABC transporter NppABCD belonging to the PepT family of transporters is implicated in the uptake of pacidamycin and other peptidyl nucleoside antibiotics, such as blastidin S, albomycin, and microcin C, across the inner membrane. These antibiotics are likely to cross the OM through the TonB-dependent receptors involved in the uptake of siderophores.

Rules of Permeation

It is likely that permeation of different classes of compounds is affected by the outer membrane barrier and by the active efflux to different degrees. At present no rules exist to predict whether increasing uptake or reducing efflux would be the most efficient way to increase the potency of a specific class of compounds. It is believed that these rules will emerge when we address a critical gap in knowledge about what physicochemical properties and specific functional groups define the permeation of compounds across cell walls of Gram-negative pathogens. Having such rules to guide medicinal chemistry efforts could potentially facilitate the discovery of novel anti-Gram negative drugs. Although the task is complex, recent success in the development of a predictive model for drug accumulation in *Caenorhabditis elegans* is inspiring. Several investigators led the efforts to develop experimental approaches and to generate representative data sets of compound penetration into difficult Gram-negative pathogens. These studies complement the activity-based approaches to identify physicochemical properties that facilitate efflux of compounds.

In addition, the first bacterial membrane models reflecting the biological complexity are emerging and could be used to address a wide range of questions about protein–lipid packing and dynamics that are not accessible by experiments. With these models and methods it is now possible to study interactions of antimicrobial peptides and antibiotics with membranes and to identify their preferable penetration routes. The "ideal" rules of permeation will also have to account for genetic and cellular processes that change the expression of efflux pumps and influx porins or modify the permeability of the OM in response to the uptake of antibiotics.

Efflux Pump Inhibitors (EPIs)

The unquestionably significant impact of multidrug efflux pumps on bacterial physiology and the resistance to antibiotics in clinical settings makes them attractive targets for inhibition. Several classes of EPIs have been reported in the literature and are being pursued in drug development programs. Phenylalanyl-arginyl-β-naphthylamide and analogous arylamines are broad-spectrum EPIs with activities against Gram-negative bacteria; arylpiperidines and more recently pyrinopyridine are inhibitors of AcrAB-TolC from *E. coli*. Likewise, pyridopyrimidines are specific inhibitors of *P. aeruginosa* MexB and *E. coli* AcrB. Importantly, broad-spectrum EPIs that target multiple efflux pumps have been reported not only to restore activities of antibiotics but also to reduce frequencies of antibiotic resistance. As with many combination therapies, the task of EPI development is not easy, and multiple hurdles must be overcome, starting with the choice of an antibiotic for potentiation and all the way to matching the pharmacological properties of an EPI/antibiotic pair. At the same time, some efflux pumps are critical for virulence and biofilm formation, for example, MexGHI-OpmH and MexEF-OprN from *P. aeruginosa*. EPIs effective against such transporters, if available, could be useful alternative therapeutics on their own.

GROWTH AND MULTIPLICATION OF BACTERIA

Bacteria divide by binary fission and cell divides to form two daughter cells. Nuclear division precedes cell division and therefore, in a growing population, many cells having two nuclear bodies can be seen. Bacterial growth may be considered as two levels, increase in the size of individual cells and increase in number of cells. Growth in numbers can be studied by bacterial counts that of total and viable counts. The total count gives the number of cells either living or not and the viable count measures the number of living cells that are capable of multiplication.

Bacterial Growth Curve

When bacteria is grown in a suitable liquid medium and incubated its growth follows a definite process. If bacterial counts are carried out at intervals after innoculation and plotted in relation to time, a growth curve is obtained. The curve shows the following phase:

- Lag phase: Immediately following innoculation there is no appreciable increase in number, though there may be an increase in the size of the cells. This initial period is the time required for adaptation to the new environment and this lag phase varies with species, nature of culture medium and temperature.

- Log or exponential phase: Following the lag phase, the cell starts dividing and their numbers increase exponentially with time.

- Stationary phase: After a period of exponential growth, cell division stops due to depletion of nutrient and accumulation of toxic products. The viable count remains stationary as an equilibrium exists between the dying cells and the newly formed cells.

- Phase of decline: This is the phase when the population decreased due to cell death.

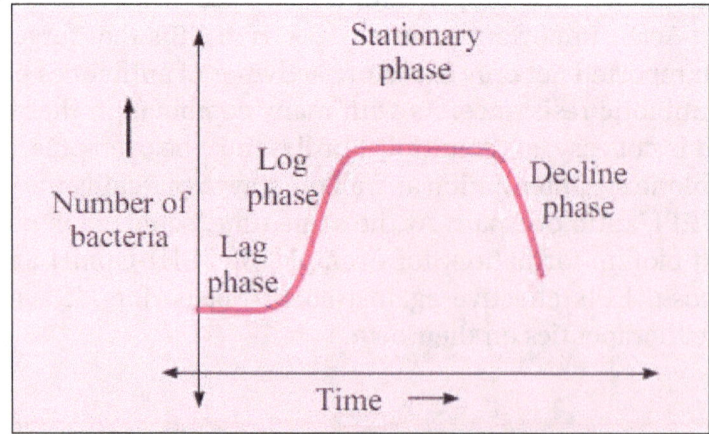

The growth curve of bacteria showing different phases.

The various stages of bacterial growth curve are associated with morphological and physiological alterations of the cells. The maximum cell size is obtained towards the end of the lag phase. In the log phase, cells are smaller and stained uniformily. In the stationary phase, cells are frequently gram variable and show irregular staining due to the presence of intracellular storage granules. Sporulation occurs at this stage. Also, many bacteria produce secondary metabolic products such as exotoxins and antibiotics. Involution forms are common in the phase of decline.

Factors that affect the Growth of Bacteria

Many factors affect the generation time of the organism like temperature, oxygen, carbon dioxide, light, pH, moisture, salt concentration.

Nutrition

The principal constituents of the cells are water, proteins, polysaccharides, lipids, nucleic acid and mucopeptides. For growth and multiplication of bacteria, the minimum nutritional requirement is water, a source of carbon, nitrogen and some inorganic salts.

Bacteria can be classified nutritionally, based on their energy requirement and on their ability to synthesise essential metabolites. Bacteria which derive their energy from sunlight are called phototrophs, those who obtain energy from chemical reactions are called chemotrophs. Bacteria which can synthesise all their organic compounds are called autotrophs and those that are unable to synthesise their own metabolites are heterotrophs.

Some bacteria require certain organic compounds in minute quantities. These are know as growth factors or bacterial vitamins. Growth factors are called essential when growth does not occur in their absence, or they are necessary for it.

Oxygen

Depending on the influence of oxygen on growth and viability, bacteria are divided into aerobes and anaerobes.

Aerobic bacteria require oxygen for growth. They may be obligate aerobes like cholera, vibrio, which will grow only in the presence of oxygen or facultative anaerobes which are ordinarily aerobic but can grow in the absence of oxygen.

Most bacterial of medical importance are facultative anaerobes. Anaerobic bacteria, such as clostridia, grow in the absence of oxygen and the obligate anaerobes may even die on exposure to oxygen. Microaerophilic bacteria are those that grow best in the presence of low oxygen tension.

Carbon Dioxide

All bacteria require small amounts of carbon dioxide for growth. This requirement is usually met by the carbon dioxide present in the atmosphere. Some bacteria like Brucella abortus require much higher levels of carbon dioxide.

Temperature

Bacteria vary in their requirement of temperature for growth. The temperature at which growth occurs best is known as the optimum temperature. Bacteria which grow best at temperatures of 25-40 °C are called mesophilic. Psychrophilic bacteria are those that grow best at temperatures below 20 °C. Another group of non pathogenic bacteria, thermophiles, grow best at high temperatures, 55-80 °C.

The lowest temperature that kills a bacterium under standard conditions in a given time is known as thermal death point.

Moisture and Drying

Water is an essential ingredient of bacterial protoplasm and hence drying is lethal to cells. The effect of drying varies in different species.

Light

Bacteria except phototrophic species grow well in the dark. They are sensitive to ultraviolet light and other radiations. Cultures die if exposed to light.

H-ion Concentration

Bacteria are sensitive to variations in pH. Each species has a pH range, above or below which it cannot survive and an optimum pH at which it grows best. Majority of pathogenic bacteria grow best at neutral or slightly alkaline pH (7.2– 7.6).

Osmotic Effect

Bacteria are more tolerant to osmotic variation than most other cells due to the mechanical strength of their cell wall. Sudden exposure to hypertonic solutions may cause osmotic withdrawal of water and shrinkage of protoplasm called plasmolysis.

Growth of bacteria is affected by many factors such as nutrition concentration and other environmental factors.

Some of the important factors affecting bacterial growth are:

- Nutrition concentration
- Temperature
- Gaseous concentration
- pH
- Ions and salt concentration
- Available water

Nutrient Concentration

- If culture media is rich in growth promoting substance, growth of bacteria occurs faster. Decrease in nutrient concentration decreases the growth rate.

- Different bacteria have different nutritional requirement.

The relationship between substrate concentration (nutrition) and growth rate is shown in figure.

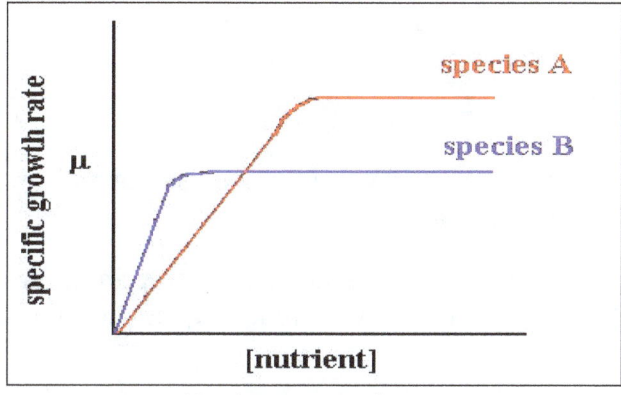

Nutrient vs. growth rate.

- With increase in concentration nutrition, growth rate of bacteria increases up to certain level and then growth rate remains constant irrespective of nutrition addition.

Temperature

- Temperature affects the growth of bacteria by various ways.

- The lowest temperature that allows the growth is called minimum temperature and the highest temperature that allows growth is called maximum temperature.

- There is no growth below minimum and above maximum temperature.

- Below minimum temperature cell membrane solidifies and become stiff to transport nutrients in to the cell, hence no growth occurs.

- Above maximum temperature, cellular proteins and enzymes denatures, so the bacterial growth ceases.

The relationship between temperature and growth rate is shown in figure below.

- When temperature is increases continuously from its minimum, growth rate of bacteria increases because the rate of metabolic reaction increases with increase in temperature.

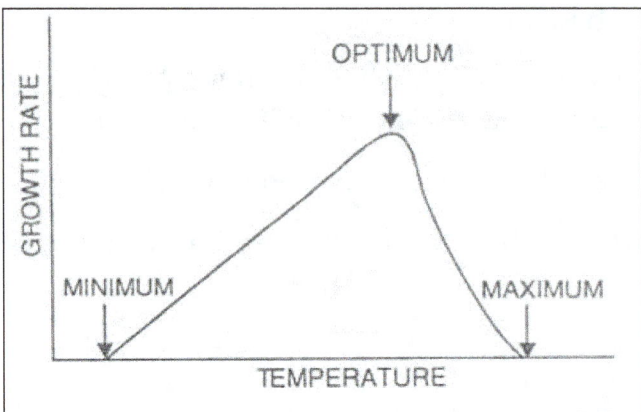

Temperature vs. growth rate.

- At certain temperature the growth rate become maximum, this temperature is known as optimal temperature.

- On further increasing the temperature above optimal, growth rate decreases abruptly and completely ceases with reaching maximum temperature.

pH

- pH affects the ionic properties of bacterial cell so it affects the growth of bacteria.

- Most of the bacteria grow at neutral pH (60.5-7.5). However there are certain bacteria that grow best at acidic or basic pH.

- Relationship between pH and bacterial growth is given in figure below.

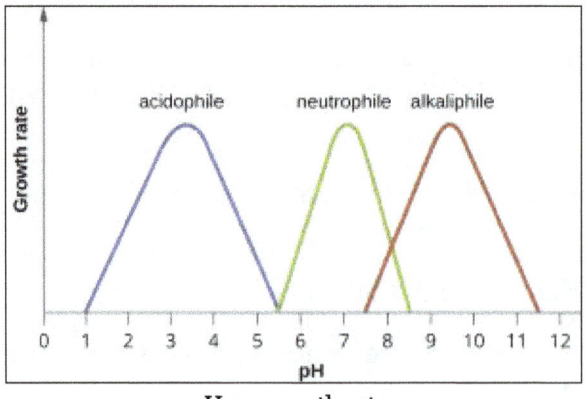

pH vs. growth rate.

Ions and Salt

- All bacteria requires metal ions such as K^+, Ca^{++}, Mg^{++}, Fe^{++}, Zn^{++}, Cu^{++}, Mn^{++} etc. to synthesize enzymes and proteins.

- Most bacteria do not require NaCl in media however they can tolerate very low concentration of salt.

- There is some halophilic bacteria such as *Archeobacteria* that require high concentration of salt in media.

Gaseous Requirement

- Oxygen and carbon-dioxide are important gases that affects the growth of bacteria.

- Oxygen is required for aerobic respiration and obligate aerobic bacteria must require O_2 for growth. eg. Mycobacterium, Bacillus.

- For obligate anaerobes Oxygen is harmful or sometime lethal. However facultative anaerobes can tolerate low concentration of O_2.

- Carbon-dioxide is needed for capnophilic bacteria. Such as Campylobacter, Helicobacter pylori.

Available Water

- Water is the most essential factor for bacterial growth.

- Available water in the culture media determines the rate of metabolic and physiological activities of bacteria.

- Sugar, salts and other substances are dissolved in water and are made available for bacteria.

Bacterial Growth Laws and their Applications

Engineering of synthetic genetic circuits holds the promise to revolutionize medical treatment and industrial production from microbes. Yet, progress over the last decade has been hindered by a lack of rational design principles to guide the interfacing of engineered components with the host organism. Recent work demonstrates that even constitutive protein expression, and much more so genetic circuits, can be strongly affected by the cell's physiological state. Thus, it appears difficult to insulate synthetic circuitry from the growth state of the host. Though orthogonal expression systems are designed to minimize this coupling, they nevertheless draw energy and resources away from the core processes in the cell.

Despite the inherent cross-talk between synthetic and endogenous elements, however, some of the resultant global effects have been shown to obey simple mathematical relations referred to as 'growth laws'. These quantitative relations may provide a framework for the design of robust synthetic systems, opening up new directions in bioengineering and biotechnology.

Empirical Growth-rate Dependence

Recent advances in the phenomenological modeling of bacterial physiology are best appreciated within a historical context. Possibly the most influential early example of coarse-grained modeling was Monod's discovery of a hyperbolic relation between the growth rate of a culture and the concentration of nutrient in the growth medium. Subject to some amendments over the years, Monod's relation continues to play an important role in current research.

The simplicity of Monod's relation belies the incredible complexity of physiological regulation in bacteria. In balanced exponential growth, however, all of this complexity operates in concert with clock-like regularity to ensure that every constituent in the cell doubles at the same rate. The seminal work by Schaechter, Maaløe and Kjeldgaard revealed that the consequences of that balance are remarkable; the macromolecular composition of Salmonella (mass of RNA, DNA, protein and cell mass itself) is largely a function of the doubling rate alone, irrespective of the detailed composition of the growth medium.

A decade after Schaechter et al.'s work, Helmstetter and Cooper carefully analyzed synchronous cultures of Escherichia coli and established that during fast growth multiple origins of replication are employed simultaneously. Coupled with the exponential dependence of the cell mass on growth rate discovered by Schaechter et al., Donachie turned these empirical observations around and postulated that both series of measurements could be explained if new rounds of DNA replication are initiated at a constant mass per origin of replication (referred to commonly as the 'Donachie mass'). Donachie's work marks one of the first times in biology that quantitative phenomenological relations were used to infer constraints on the underlying molecular mechanisms. Elucidating the details of DNA-replication initiation remains an active area of research, although any proposed mechanism must be consistent with Donachie's observation.

In the wake of these early studies, the growth rate dependence of a large catalogue of physiological parameters was measured in E. coli, driven by the meticulous efforts of Hans Bremer and coworkers. That accumulated data laid the foundations for the quantitative models of bacterial physiology that began to emerge in the late 1970's and early 1980's.

Quantitative Models of Bacterial Physiology

Growth-rate dependent physiological parameters collected over the intervening decades served as input to predictive, integrative models. Ehrenberg and Kurland used a detailed model of bacterial physiology, subject to the maximization of growth rate, as a means to quantify the costs associated with accuracy in protein translation. Growth rate maximization, along with a range of other objectives, continues to be used in modern constraint-based reconstruction and analysis of metabolic networks. In some cases, the resulting analyses have met with remarkable success in predicting changes in metabolism incurred during evolutionary adaptation, and in qualitatively addressing shifts to inefficient metabolic strategies during rapid growth. Recently, Tadmor and Tlusty used a coarse-grained model of macromolecular synthesis in E.coli to predict the growth defect associated with the deletion of ribosomal RNA operons, and suggested that molecular crowding effects set the limit on the optimum copy number of the ribosomal RNA operons. Klumpp and Hwa, focusing on transcription, developed a model for the free RNA polymerase concentration to explain the growth-rate dependence of promoter activities, and were thereby able to constrain admissible transcriptional control functions in the wake of amino acid starvation.

Although each of these studies has brought us closer to a more complete understanding of the complex interplay between regulation and physiology, the limitation of these approaches is that even the coarse-grained models involve a number of parameters that must be independently measured, for each growth condition of interest. As Klumpp et al. showed, simple constitutive gene expression in exponentially growing bacteria is strongly dependent upon the growth rate of the culture. The growth rate dependence can be traced back to changes in global parameters such as the basal transcriptional and translational rates, gene dose, and cell volume at different growth rates. By analyzing these global parameters, the authors were able to make accurate, quantitative predictions regarding the coupling between gene expression and cell growth for a number of common genetic network motifs.

The success of these 'bottom-up' models suggests that the immense complexity of genetic and metabolic regulation under different growth conditions can be captured by a limited number of growth-rate dependent global parameters (e.g., those describing transcription and translation), at least as far as gene expression is concerned. Nevertheless, such approaches are not able to explain the origin of these growth-rate dependences. Furthermore, the growth rate can be altered in many ways: in a continuous culture, it is the quality of the growth-limiting nutrient that is adjusted through the dilution rate. In batch culture, most often it is the quality of the saturating amount of nutrient that is changed. The growth rate can, of course, be modulated in many other ways, including temperature, osmolarity, antibiotics, toxin, or other conditions. In that case, a `bottom-up' approach requires careful wholesale measurements to be repeated for the dozen or so global parameters used in the model. A different perspective is needed to explain the origins of the growth rate dependence of gene expression, and at the same time allow straightforward extension to other modes of growth modulation.

Bacterial Growth Laws

Hidden within the experimental results of Schaechter et al. is the remarkable linear relation between RNA/total protein ratio and the growth rate over moderate to fast growth rates (faster than 2 hrs per doubling). In E. coli, the total RNA is approximately 85% ribosomal RNA (a fraction that is growth-rate independent over moderate to fast growth rates), and so the RNA/protein ratio is directly proportional to the mass fraction of ribosomes in the cell. Later experiments by Neidhardt and Magasanik brought this relation between ribosome content and growth rate to center stage. The linearity can be understood as a consequence of mass balance: protein mass accumulation is generated by elongating ribosomes, and in balanced exponential growth that implies a linear relation between the ribosomal mass fraction (proportional to the RNA/protein ratio) and the growth rate. The constant of proportionality ($1/\kappa t$) is then given by the reciprocal of the translational elongation rate. This interpretation of the linear relationship between ribosomal content and growth rate under changes in nutrient quality suggests a conjugate relation may be revealed by changing the translation rate.

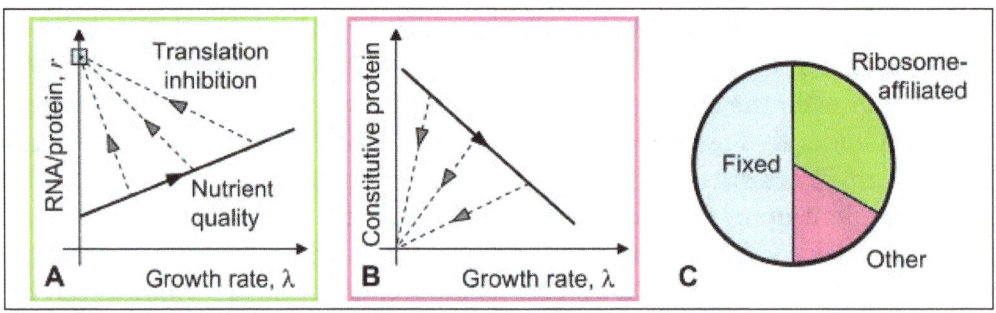

Bacterial growth laws.

In the above figure, A. When growth is modulated by changes in nutrient quality, the RNA mass fraction r (proportional to the ribosomal content) of *E. coli* increases linearly with growth rate γ (solid line): $r = r_0 + \gamma/\kappa_t$, where the parameter κ_t is related to the translation rate and r_0 is the offset. When growth is modulated by changes in translational efficiency, a conjugate relation is observed. The RNA mass fraction is inversely related to growth rate (dashed lines): $r = r_{max} - \gamma/\kappa_n$, where the parameter κ_n describes the nutrient quality of the growth medium, and r_{max} is the maximum allocation to ribosomal synthesis in the limit of complete translational inhibition. B. Symmetric linear relations are observed in the mass fraction of a constitutively expressed protein, implying a linear constraint between ribosome-affiliated and constitutive proteins. C. The simplest constraint is a three-component partition of the proteome: a fixed fraction that is invariant to growth-rate change (blue), ribosome and ribosome-affiliated proteins (green) and the remainder (pink), including constitutive proteins. For *E. coli* K-12 MG1655, the fixed fraction appears to occupy roughly half of the protein fraction.

When translation is inhibited, either by subinhibitory levels of antibiotic or targeted mutations of the ribosomal proteins, a second linear relation between the RNA/protein ratio and the growth rate becomes apparent. In this case, the constant of proportionality ($1/\kappa_n$) is inversely related to the nutrient quality. More remarkably, the two linear relations describing the RNA mass fraction observed under changes in nutrient quality and translational ability are reflected almost perfectly in the mass fraction of a constitutively expressed protein. Although these empirical relations are independent of any interpretation, the mirror-symmetry between ribosome-affiliated and constitutive proteins suggests the existence of a linear constraint. The empirical relations shown in figures can be unified into a phenomenological theory of bacterial growth (Box 1) by postulating a minimal three-component partitioning of the proteome that includes a growth-rate invariant fraction ϕ_{fixed}, a fraction containing ribosome and ribosome-affiliated proteins, and a third fraction containing the remainder (including constitutive proteins).

Taken together, these results suggest that the origin of the growth rate dependence of constitutive gene expression arises from a tug-of-war between the need for protein synthesis (mediated by ribosomal proteins) and nutrient uptake/processing (mediated by other non-ribosomal proteins). Qualitatively, this type of flux balance is expected in any autocatalytic (self-reproducing) reaction scheme, as was first proposed by Hinshelwood. Koch applied Hinshelwood's analysis to study the biophysical constraints on bacterial growth-rate regulation by considering an autocatalytic loop composed of ribosomal and non-ribosomal proteins. A similar two-component model was used by Alon and coworkers in their recent study of promoter activities under various modes of growth inhibition.

One limitation of a two-component proteome partition is that under favourable nutrient conditions or maximum translational inhibition, the predicted ribosomal protein fraction would be close to one. In fact, including all auxiliary proteins required for translation (initiation factors, elongation factors, etc.), that fraction is substantially less than one. The minimum partition model, then, must include at least three fractions to account for this disparity. Quantitative characterization of the two distinct linear relations between the ribosomal proteins and the growth rate, along with the explicit recognition of the fixed fraction, makes it possible to expand Koch's analysis into a predictive theory. As with Donachie's mass, this phenomenological model serves as a constraint for proposed molecular mechanisms of the underlying regulatory processes.

Of more direct relevance to biotechnology and synthetic biology applications, the growth theory outlined in figure provides a conceptual framework to guide the interfacing between synthetic constructs and the host organism. Combination of the two linear relations and the constraint on the total mass fraction yields a mathematical expression identical to Kirchoff's laws applied to two resistors in series. In this formulation, increase in translation rate is characterized by an increased conductance in the protein synthesis branch (increased κ_t), whereas an increase in nutrient quality is characterized by an increased conductance in the nutrient branch (increased κ_n).

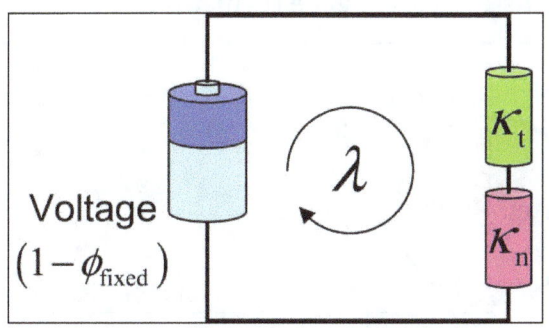

Analogy with Kirchoff's law.

In the above figure, the Monod-like relation for growth, Eq. $\lambda = \lambda_{max} \dfrac{\kappa_n}{\kappa_n + \kappa_t}$, is mathematically identical to the description of electric current flow through a pair of resistors connected in series to a battery with voltage $(1 - \varphi_{fixed})$. In this analogy, the growth rate γ is the current through the resistors. The translation- and nutrient- modes of growth limitation correspond to changing the conductance of one of the resistors, while the expression of unnecessary protein corresponds to changing the applied voltage by increasing φ_{fixed}.

Applications of the Growth Laws

The analogy to Kirchoff's laws is significant in so far as it allows the growth theory to be extended to different modes of growth rate modulation, particularly to the growth defect associated with heterologous protein expression. Heterologous protein expression effectively expands the fixed protein fraction $\varphi fixed$, which in the Kirchoff's law analogy corresponds to a reduction of the driving voltage. Consistent with this interpretation, overexpression of an unnecessary protein in E. coli results in a linear decrease in the growth rate, with the zero-growth limit occurring when the overexpressed protein occupies a mass fraction of $(1 - \varphi fixed)$, in agreement with results by others using different expression vectors and proteins. These results disagree with those of Dekel and Alon, who reported a quadratic reduction in growth rate upon protein overexpression from the lac operon. Closer inspection shows that the protein expression level, and the concomitant growth rate reduction, reported in was over a comparatively narrow range ($< 0.5\%$ of the proteome), whereas the effects described in Refs. covered expression levels up to $30\text{–}40\%$ of the proteome.

With a large part of the proteome occupied by the fixed fraction and the overexpressed protein, less of the proteome is available for the remaining fractions responsible for translation and nutrient uptake/processing. The observed destruction of ribosomes upon overexpression may simply reflect the native response to nutrient limitation. One immediate conclusion is that the metabolic load associated with overexpression can be mitigated by reducing the nominal fixed protein fraction φ_{fixed}, which may lead to alternate methods of increasing heterologous protein yield that augment existing strategies of strain-optimization. Of wider interest in evolutionary studies, the relative growth defect associated with overexpression (often called the "fitness cost") provides a basis for quantifying different strategies of gene regulation.

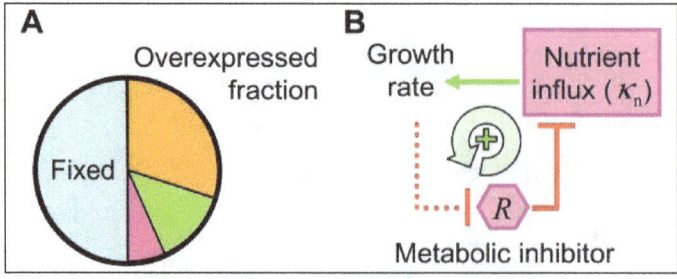

Some applications of the growth laws.

In the above figure, A. The burden of protein overexpression. Expression of an unnecessary protein (orange) effectively decreases the fraction allocable to the protein sectors responsible for protein synthesis (green) and nutrient uptake processing (pink), leading to a decrease in the growth rate. B. Growth-mediated feedback. Constitutive expression of a toxin affecting nutrient influx (R) could lead to bistability through positive feedback generated by the interdependence of gene expression levels and growth rate (dotted line). A decrease in growth rate under conditions of nutrient limitation results in an increase in the constitutively expressed toxin R, reinforcing further growth rate reduction.

Cross-talk between endogenous and synthetic elements has long frustrated the advance of synthetic biology, despite the growing list of well-characterized components. The phenomenological framework emerging from recent work on E. coli not only suggests how this cross-talk can be incorporated into network design, but also offers a strategy to exploit coupling to host physiology for rational purposes. The interdependence of gene expression and growth rate can lead to global feedback loops, as suggested by Narang and co-workers in a series of critical analyses of the regulation underlying carbon utilization in E. coli, though the existence of different modes of growth inhibition makes the feedback scenarios much richer than the dilution models considered in that work.

It should, for example, be possible to generate complicated phenotypic switching through growth-mediated feedback if the cell expresses a toxin inhibiting nutrient influx. Although self-inflicted growth-arrest seems counterintuitive to the survival of the organism, quiescent cells are resistant to many antibiotics and growth-arrest represents a bet-hedging strategy to ensure long-term viability of the colony (a phenomenon called 'bacterial persistence'). Lou et al. proposed a growth-mediated positive feedback mechanism to facilitate spontaneous growth transition to the dormant state using a dilution model with highly-cooperative regulatory interactions controlling the expression of a toxin. Klumpp et al. further pointed out that growth-dependent global feedback could lead to persistence for constitutively expressed toxins.

In an industrial setting, bistability resulting from global feedback effects may arise unintentionally due to the action of the synthetic network used to drive heterologous expression. Regulatory motifs used in expression vectors must then be optimized to avoid bistability to prevent overgrowth of the population by the low-producing phenotype. Bistability of this type has already been reported by You and coworkers, where the growth defect associated with the overexpression of a foreign polymerase generates the requisite feedback. In a therapeutic setting, phenotypic bistabiltiy mediated by overexpression may underlie expression of motility and virulence factors in some bacterial pathogens.

INTERACTIONS WITH OTHER ORGANISMS

Despite their apparent simplicity, bacteria can form complex associations with other organisms. These symbiotic associations can be divided into parasitism, mutualism and commensalism. Due to their small size, commensal bacteria are ubiquitous and grow on animals and plants exactly as they will grow on any other surface. However, their growth can be increased by warmth and sweat, and large populations of these organisms in humans are the cause of body odour.

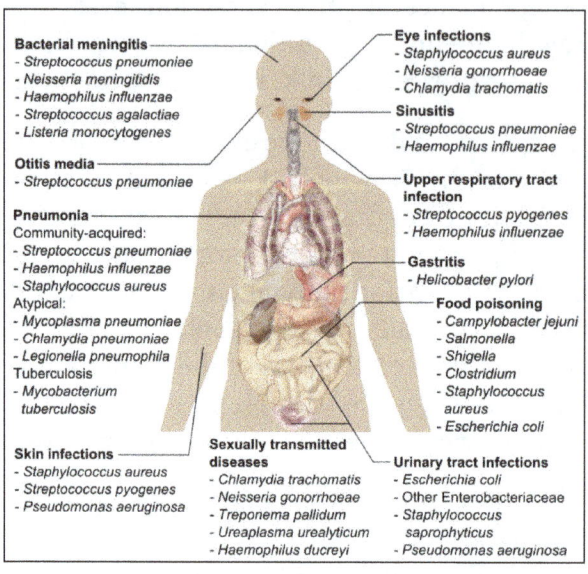

Bacterial infections and main species involved.

Predators

Some species of bacteria kill and then consume other microorganisms, these species are called predatory bacteria. These include organisms such as Myxococcus xanthus, which forms swarms of cells that kill and digest any bacteria they encounter. Other bacterial predators either attach to their prey in order to digest them and absorb nutrients, such as Vampirovibrio chlorellavorus, or invade another cell and multiply inside the cytosol, such as Daptobacter. These predatory bacteria are thought to have evolved from saprophages that consumed dead microorganisms, through adaptations that allowed them to entrap and kill other organisms.

Mutualists

Certain bacteria form close spatial associations that are essential for their survival. One such mutualistic association, called interspecies hydrogen transfer, occurs between clusters of anaerobic bacteria that consume organic acids, such as butyric acid or propionic acid, and produce hydrogen, and methanogenic Archaea that consume hydrogen. The bacteria in this association are unable to consume the organic acids as this reaction produces hydrogen that accumulates in their surroundings. Only the intimate association with the hydrogen-consuming Archaea keeps the hydrogen concentration low enough to allow the bacteria to grow.

In soil, microorganisms that reside in the rhizosphere (a zone that includes the root surface and the soil that adheres to the root after gentle shaking) carry out nitrogen fixation, converting nitrogen gas to nitrogenous compounds. This serves to provide an easily absorbable form of nitrogen for many plants, which cannot fix nitrogen themselves. Many other bacteria are found as symbionts in humans and other organisms. For example, the presence of over 1,000 bacterial species in the normal human gut flora of the intestines can contribute to gut immunity, synthesise vitamins, such as folic acid, vitamin K and biotin, convert sugars to lactic acid, as well as fermenting complex undigestible carbohydrates. The presence of this gut flora also inhibits the growth of potentially pathogenic bacteria (usually through competitive exclusion) and these beneficial bacteria are consequently sold as probiotic dietary supplements.

Pathogens

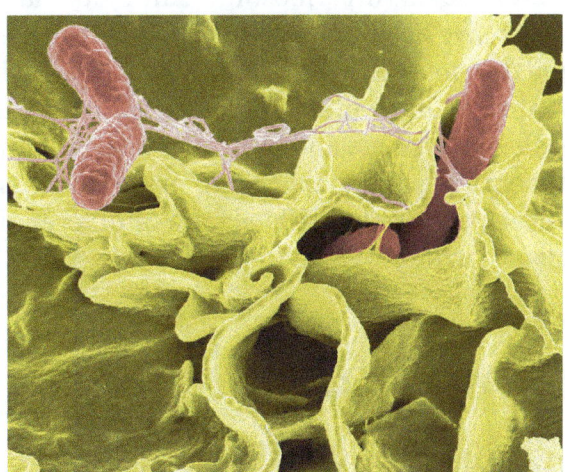

Colour-enhanced scanning electron micrograph showing Salmonella
typhimurium (red) invading cultured human cells.

If bacteria form a parasitic association with other organisms, they are classed as pathogens. Pathogenic bacteria are a major cause of human death and disease and cause infections such as tetanus, typhoid fever, diphtheria, syphilis, cholera, foodborne illness, leprosy and tuberculosis. A pathogenic cause for a known medical disease may only be discovered many years after, as was the case with Helicobacter pylori and peptic ulcer disease. Bacterial diseases are also important in agriculture, with bacteria causing leaf spot, fire blight and wilts in plants, as well as Johne's disease, mastitis, salmonella and anthrax in farm animals.

Each species of pathogen has a characteristic spectrum of interactions with its human hosts. Some organisms, such as Staphylococcus or Streptococcus, can cause skin infections, pneumonia, meningitis and even overwhelming sepsis, a systemic inflammatory response producing shock, massive vasodilation and death. Yet these organisms are also part of the normal human flora and usually exist on the skin or in the nose without causing any disease at all. Other organisms invariably cause disease in humans, such as the Rickettsia, which are obligate intracellular parasites able to grow and reproduce only within the cells of other organisms. One species of Rickettsia causes typhus, while another causes Rocky Mountain spotted fever. Chlamydia, another phylum of obligate intracellular parasites, contains species that can cause pneumonia, or urinary tract infection and may be involved in coronary heart disease. Finally, some species, such as Pseudomonas aeruginosa, Burkholderia cenocepacia, and Mycobacterium avium, are opportunistic pathogens and cause disease mainly in people suffering from immunosuppression or cystic fibrosis.

Bacterial infections may be treated with antibiotics, which are classified as bacteriocidal if they kill bacteria, or bacteriostatic if they just prevent bacterial growth. There are many types of antibiotics and each class inhibits a process that is different in the pathogen from that found in the host. An example of how antibiotics produce selective toxicity are chloramphenicol and puromycin, which inhibit the bacterial ribosome, but not the structurally different eukaryotic ribosome. Antibiotics are used both in treating human disease and in intensive farming to promote animal growth, where they may be contributing to the rapid development of antibiotic resistance in bacterial populations. Infections can be prevented by antiseptic measures such as sterilising the skin prior to piercing it with the needle of a syringe, and by proper care of indwelling catheters. Surgical and dental instruments are also sterilised to prevent contamination by bacteria. Disinfectants such as bleach are used to kill bacteria or other pathogens on surfaces to prevent contamination and further reduce the risk of infection.

BACTERIAL MORPHOLOGY

Bacteria are very small unicellular microorganisms ubiquitous in nature. They are micrometres ($1\mu m = 10^{-6}$ m) in size. They have cell walls composed of peptidoglycan and reproduce by binary fission. Bacteria vary in their morphological features.

The Most Common Morphologies are:

Coccus (Plural – Cocci)

Spherical bacteria; may occur in pairs (diplococci), in groups of four (tetracocci), in grape-like clusters (Staphylococci), in chains (Streptococci) or in cubical arrangements of eight or more (sarcinae).

For example – Staphylococcus aureus, Streptococcus pyogenes.

Bacillus (Plural – Bacilli)

Rod-shaped bacteria; generally occur singly, but may occasionally be found in pairs (diplo-bacilli) or chains (streptobacilli).

For example – Bacillus cereus, Clostridium tetani.

Spirillum (Plural – Spirilla)

Spiral-shaped bacteria.

For example – Spirillum, Vibrio, Spirochete species.

Some Bacteria have other Shapes such as

Coccobacilli – Elongated spherical or ovoid form.

Filamentous – Bacilli that occur in long chains or threads.

Fusiform – Bacilli with tapered ends.

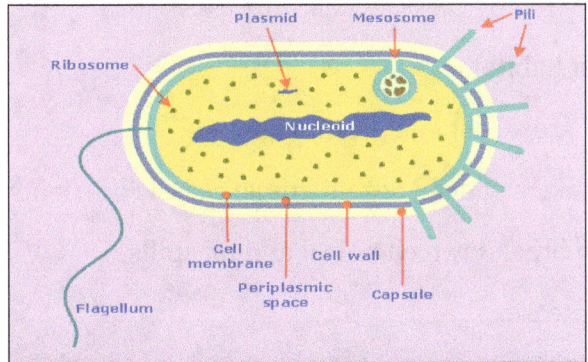

Structure and contents of a typical gram positive bacteria cell.

- Most numerous organisms on earth.

- Earliest life forms (fossils date 2.5 billion years old).

- Microscopic prokaryotes (no nucleus non membrane-bound organelles).

- Contain ribosomes.

- Infoldings of the cell membrane carry on photosynthesis and respiration.

- Surrounded by protective cell wall containing peptidoglycan (protein- carbohydrate).

- Many are surrounded by a sticky, protective coating of sugars called the capsule or glycocalyx (can attach to other bacteria or host).

- Have only one circular chromosome.

- Have small rings of DNA called plasmids.

- May have short, hair like projections called pili on cell wall to attach to host or another bacteria when transferring genetic material.

- Most are unicellular.

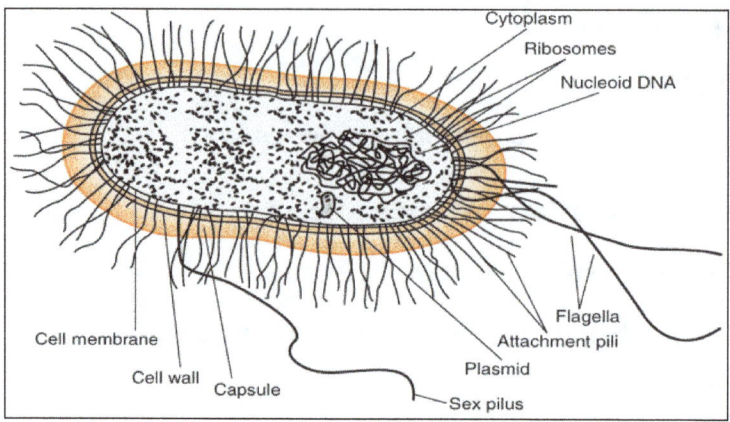

A diagram of bacteria.

- Found in most habitats.

- Most bacteria grow best at a pH of 6.5 to 7.0.

- Main decomposers of dead organisms so recycle nutrients.

- Some bacteria breakdown chemical and oil spills.

- Some cause disease.

- Move by flagella.

- Some can form protective endospores around the DNA when conditions become unfavorable; may stay inactive several years and then re-activate when conditions favorable.

- Classified by their structure, motility (ability to move), molecular composition, and reaction to stains (Gram stain).

- Grouped into 2 kingdoms – Eubacteria (true bacteria) and Archaebacteria (ancient bacteria).

- Once grouped together in the kingdom Monera.

- Classified by their structure, motility (ability to move), molecular composition, and reaction to stains (Gram stain).

Structure	Function
Cell Wall	Protect the cell and gives shape.
Outer Membrane	Protect the cell against some antibiotics (only present in Gram negative cells).
Cell Membrane	Regulates movements of materials into and out of the cell; contains enzymes important to cellular respiration.

Cytoplasm	Contains DNA, ribosomes, and organics compounds required to carry out life processes.
Chromosome	Carries genetic information inherited from past generations
Plasmid	Contains some genes obtain through genetic recombination.
Capsule, and slime layer	Protect the cell and assist in attaching the cell to other surfaces.
Endospore	Protects the cell against harsh environments conditions, such as heat or drought.
Pilus (Pili)	Assist the cell in attaching to others surfaces, which is important for genetic recombination.
Flagellum	Moves the cell.

Kingdom Archaebacteria

- Found in harsh environments (undersea volcanic vents, acidic hot springs, salty water).

- Cell walls without peptidoglycan.

- Subdivided into 3 groups based on their habitat — methanogens, thermoacidophiles, and extreme halophiles.

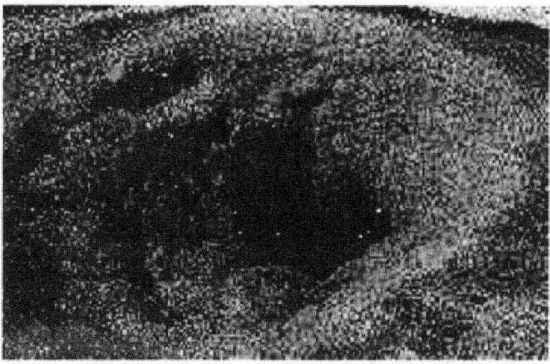
Methanogens.

Methanogens

- Live in anaerobic environments (no oxygen).

- Obtain energy by changing H_2 and CO_2 gas into methane gas.

- Found in swamps, marshes, sewage treatment plants, digestive tracts of animals.

- Break down cellulose for herbivores (cows).

- Produce marsh gas or intestinal gas (methane).

Extreme Halophiles

- Live in very salty water.

- Found in the Dead Sea, Great Salt Lake, etc.

- Use salt to help generate ATP (energy).

Thermoacidophiles (Thermophiles)

- Live in extremely hot (110 °C) and acidic (pH 2) water.

- Found in hot springs in Yellowstone National Park, in volcanic vents on land, and in cracks on the ocean floor that leak scalding acidic water.

Kingdom Eubacteria

- Most bacteria in this kingdom.

- Come in 3 basic shapes — cocci (spheres), bacilli (rod shaped), spirilla (cork-screw shape).

- Bacteria can occur in pairs (diplo – bacilli or cocci).

- Bacteria occurring in chains are called strepto – bacilli or cocci.

- Bacteria in grapelike clusters are called staphylococci.

- Most are heterotrophic (can't make their own food).

- Can be aerobic (require oxygen) or anaerobic (don't need oxygen).

- Subdivided into 4 phyla – Cyanobacteria (blue-green bacteria), Spirochetes, Gram-positive, and Proteobacteria.

- Can be identified by Gram staining (gram positive or gram negative).

Phylum Cyanobacteria

- Gram negative.
- Carry on photosynthesis and make oxygen.
- Called blue-green bacteria.
- Contain pigments called phycocyanin (red and blue) and chlorophyll a (green).
- May be red, yellow, green, brown, black, or blue-green.
- Some grow in chains (e.g. Oscillatoria) and have specialized cells called hetero-cyst that fix nitrogen.

- First bacteria to re-enter devastated areas.

- Anabaena that lives on nitrates and phosphates in water can overpopulate and cause "population blooms" or eutrophication.

- After eutrophication, the cyanobacteria die, decompose, and use up all the oxygen for fish.

Phylum Spirochetes

- Gram positive.

- Have flagella at each end so move in a corkscrew motion.

- Some are aerobic (require oxygen); others are anaerobic.

- May be free-living, parasitic, or live symbiotically with another organism.

Phylum Proteobacteria

- Largest and most diverse bacterial group.

- Subdivided into Enteric bacteria, Chemoautotrophic bacteria, & Nitrogen- fixing bacteria.

Enteric Bacteria

- Gram negative heterotrophs.

- Can live in aerobic and anaerobic environments.

- Includes E. coli that lives in the intestinal tract making vitamin K and helping break down food.

- Salmonella causes food poisoning.

Chemoautotrophs

- Gram negative bacteria that obtain energy from minerals.

- Iron-oxidizing bacteria found in freshwater ponds use iron salts for energy.

Nitrogen-Fixing Bacteria

- Rhizobium is Gram negative and live in legume root nodules.

- 80% of atmosphere is N_2, but plants can't use nitrogen gas.

- Nitrogen-fixing bacteria change N_2 into usable ammonia (NH_3).

- Important part of the Earth's nitrogen cycle.

MECHANICAL WORLD OF BACTERIA

Bacteria occupy a broad variety of ecological niches on Earth. Their long evolutionary history has exposed them to vastly different environments, and they have evolved remarkable plasticity in response to locally changing physicochemical conditions. In particular, bacteria can detect and respond to chemical, thermal, and mechanical cues, as well as to electric and magnetic fields. How do these cues influence bacterial behaviors in natural environments? Characterizing bacterial behavior in realistic contexts requires integrating a spectrum of environmental stimuli to which they respond, and doing so in physical configurations representative of their natural habitats. Such analyses are critical to comprehensively understand bacterial biology and to thereby make progress in promoting or restricting bacterial growth in medical, industrial, and agricultural realms.

Mechanics is an integral part of eukaryotic cell biology: numerous studies have demonstrated the importance of fluid flow and surface mechanics in mammalian cell growth and behavior at many different length scales. In contrast, microbiology has traditionally focused on the influence of the chemical environment on bacterial behavior. Hence, for decades, growth in well-mixed batch cultures and on agar plates were the methods of choice for studies of bacterial physiology. As a result, the community has only recently recognized that mechanics also play a significant role in microbial biology on surfaces: fluid flow and contact between cells and surfaces are two ubiquitous and influential features of bacterial existence in natural environments. Advances in microscale engineering and microscopy now provide us with powerful tools to explore, at the relevant spatial scales, the roles physical forces play in bacterial sensory perception and adaptation. These new experimental platforms have revealed that bacteria are attuned to mechanical forces and, indeed, can exploit mechanics to drive adaptive behavior.

Swimming motility provides an elegant example of how bacteria are influenced by the mechanical nature of their surroundings. As a consequence of their small size (~1 μm), bacteria live in environments dominated by viscosity, which stands in contrast to the meter-scale world of humans in which dynamics are dominated by inertia. Fluid motion can be broadly characterized by the Reynolds number (Re), which compares the magnitudes of inertial forces and viscous forces in a given flow ($Re = \rho UL/\mu$ where U is a typical fluid speed, L a typical length scale, ρ the density of the fluid and μ its viscosity). We humans live a high Reynolds number life (at least 10^4), as we are meter-scale organisms moving at speeds on the order of meters per second. But swimming microorganisms live at Reynolds numbers far below unity (at most 10^{-3}). To self-propel in such a regime, bacteria use motorized flagella that convert mechanical actuation (rotation) into net displacement. Thus, many bacteria have evolved a biological machine - the flagellum and its associated motor - to adapt to the mechanical properties of their (purely viscous) environment.

Outside of the oceans, most bacteria in nature exist on surfaces, rather than in the bulk liquid of their fluid environments. Bacteria are equipped to live at the liquid-solid

interface via the secretion of adhesive structures such as flagella, pili, exopolysaccha-
rides, and other matrix components. The mechanical environment of surface-associated
bacteria is remarkably different than that of their free-floating counterparts. From initial contact, a surface-attached bacterium will experience a local force that is normal to
the surface, usually referred to as an adhesive force. In an environment with flow, the
viscosity of the surrounding fluid generates a hydrodynamic (shear) force on the cell
that is tangential to the surface in the direction of the flow. Surface motility may produce
a friction force that is tangential to the cell wall and localized at the interface with the
substrate. The principles of mechanics dictate that the forces on a stationary or steadily
moving cell must balance, so that a local adhesive force toward the substrate at one point
on the cell must be balanced by repulsive forces due to compression elsewhere.

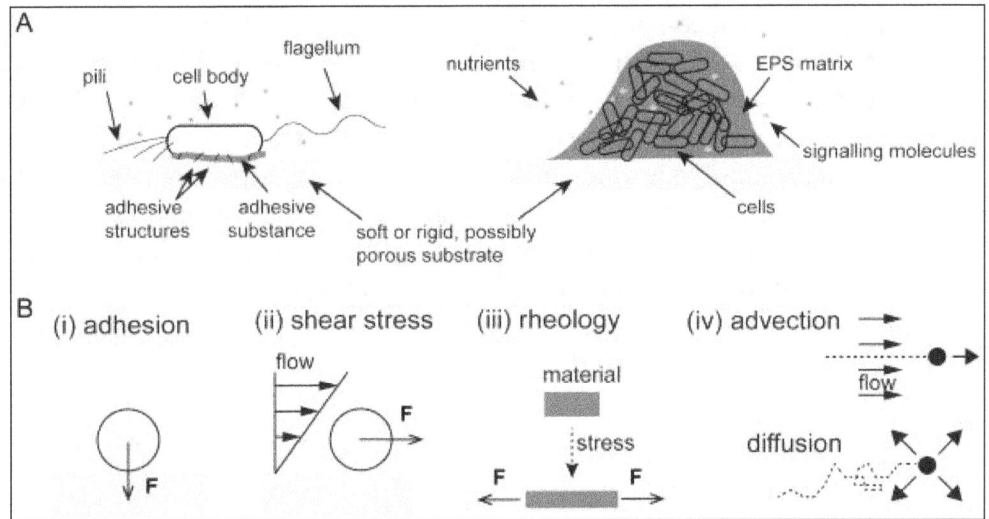

Bacteria experience a variety of mechanical effects on surfaces.

In the above figure, it is shown, (A) Flagella, pili, and adhesive substances are useful for
attachment of individual bacterial cells to surfaces. Extracellular polymeric substances
(EPS) aid in maintaining the integrity of community structures composed of multiple
cells. Bacteria use diffusible signaling molecules, chemical weapons, and soluble public
goods to interact within such communities. (B) A cell attaching to a surface is subject
to a local adhesive force F in the direction normal to the surface. Shear stresses due to
fluid flow generate a force F on the cell that is parallel to the surface. Bacteria experience the rheological properties of their surrounding extracellular matrix, which flows
and deforms upon application of forces. Fluid flow (advection) and Brownian motion
(diffusion) transport soluble compounds (black dots) that are released and/or internalized by bacteria.

Surface-attached bacterial cells can multiply to form large groups that develop into
organized communities termed biofilms. At this multicellular scale, additional mechanical effects become relevant. Attachment of a cell to a surface induces secretion
of a mixture of proteins, polysaccharides, and DNA that form a surrounding matrix

(EPS; extracellular polymeric substances) with both viscous and elastic properties. These extracellular polymers bind surface-attached cells and their progeny together in biofilm communities, and the rheology of the secreted matrix likely has important implications for the growth, spatial arrangement, and resilience of the resulting multicellular structures. The spatiotemporal distribution of small molecules internalized and/or released by bacterial cells residing within these communities can be strongly affected by the flow environment that the community experiences, with substantial and distinct consequences for individual and collective behaviors.

Here, we highlight how these mechanical effects play roles in bacterial behavior at the level of single cells and of multicellular structures. We discuss strategies that bacterial cells deploy specifically on surfaces, including enhanced adhesion under fluid flow, exploration *via* surface-specific motility, and control of cell shape to enhance colonization. At the level of multicellular structures, we discuss how the rheology of polymeric matrices affects populations growing in biofilms and how flow influences these structures. We also describe how fluid flow affects the transport of small molecules used in social interactions, e.g. quorum sensing, between individual bacterial cells. Finally, we provide insight into the scalability of the effects of mechanics on bacteria, i.e. how phenomena at the level of single cells influence emergent collective behaviors and group fitness consequences in multicellular communities.

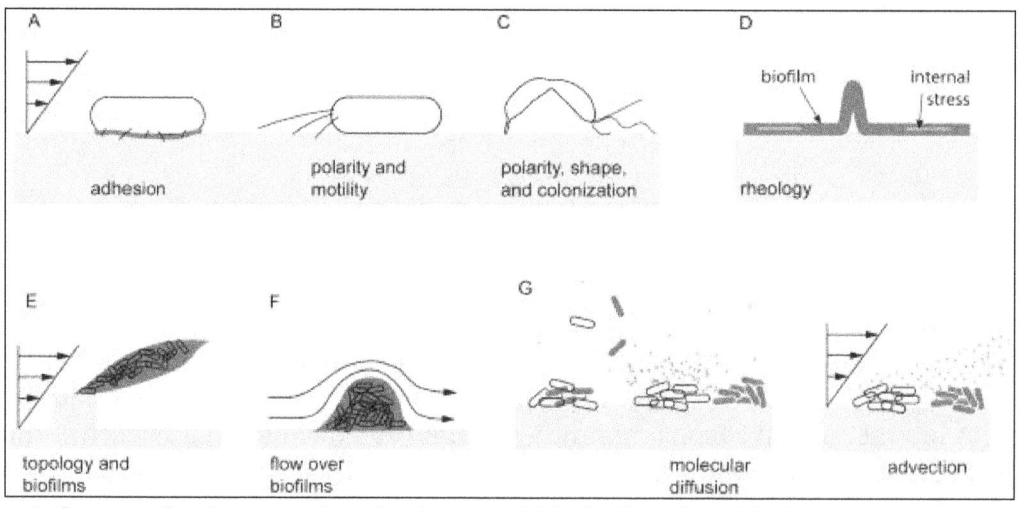

Influences of environmental mechanics on individual cells and multicellular communities.

In the above figure, (A) On surfaces and in the presence of flow, individual bacterial cells use short appendages (fimbriae) and other adhesive structures or substances to remain strongly attached, and (B) they use motorized pili localized at their poles to move against the flow. (C) Bacteria exploit their shapes to orient in flow thereby enhancing surface colonization. (D) At the scale of multicellular communities, bacteria can alter the rheology of the extracellular polymeric substances (EPS) to optimize growth. (E, F) Flow modifies the architecture of bacterial biofilms by driving formation of filamentous structures called streamers, which can obstruct flow but also capture cells and metabolites suspended in the surrounding fluid. (G) Transport of nutrients and other solutes by diffusion and advection drives the growth of and interactions between surface-associated bacteria.

Mechanics at the Level of Single Cells

To initiate and maintain intimate contact with solid surfaces, bacteria leverage a wide variety of adhesion strategies. Many bacteria, upon attaching to a surface, will secrete a mixture of EPS, which increases their affinity for porous, rough, and chemically heterogeneous surfaces. Bacteria also construct protein structures on their exteriors that enhance their adhesion to surfaces. For example, appendages such as pili and fimbriae aid cells in overcoming repulsive forces between the cell membrane and abiotic surfaces. EPS secretion and pilus formation are active areas of investigation and have been reviewed elsewhere. Previous reviews have also highlighted surface-specific motility such as swarming and twitching. Here we focus on strategies bacteria use – often employing fimbriae, pili and EPS –to maintain attachment to surfaces or to optimize surface colonization in flow environments.

Holding Tight

The shear stress generated by flow at a solid-liquid interface can easily overcome the adhesive forces anchoring cells onto a surface, potentially detaching them from substrata. In flow, a cell experiences a drag force that is well-estimated as $F_{drag} = A\sigma_s$, where A is the area of the cell exposed to flow (approximately the product of length and width for a rod-shape bacterium) and σ_s is the local shear stress at the surface. In a microfluidic channel with rectangular cross section of height h and width w, and given flow rate Q (in m^3 per second), the shear stress at the center of the wall can be estimated by $\sigma_s = 6Q\mu/wh^2$ where μ is the fluid viscosity. While shear stress depends highly on the geometry of the flow, it is generally larger in environments with higher flow speeds. Thus, the drag force on an attached cell typically increases with flow intensity, and the attachment strength required for a cell to resist removal by shear will depend on the flows that characterize its environmental niche. Cell adhesion forces range from a few to hundreds of picoNewtons (pN), which is sufficient to maintain attachment in a variety of flow environments. These forces also strongly depend on chemistry and mechanical properties of the substrate.

Some bacteria, like the prosthecate *Caulobacter crescentus*, stand out among microbial models of shear resistance with their extremely strong surface attachment. Single *C. crescentus* cells construct an adhesive holdfast, which is composed of a sticky substance that localizes at the cell poles, to withstand forces as large as 1 μN, which effectively renders them irreversibly surface-attached. C. crescentus cells can withstand shear stresses as high as 1 MPa (their typical surface area being on the order of 1 μm^2). It is not clear why C. crescentus evolved such extreme attachment strength, given that the typical shear stress in their natural freshwater environments is expected to be orders of magnitude lower. One hypothesis is that strong attachment prevents grazing by predators.

Catch Bonds

Paradoxically, multiple examples exist in which increasing shear stress enhances cell attachment to surfaces. For example, *Escherichia coli* is subject to flows spanning a wide range of intensities as it colonizes different host tissues, and it has evolved adaptable fimbriae that counteract removal by flow to optimize colonization in these diverse environments. Typical bacterial fimbriae fail to maintain adherence upon application of a sufficiently large force, whereas among many strains of *E. coli*, type I fimbriae attachment is enhanced under increasing tensile load. In these cases, type I fimbriae are capped with a tip protein called FimH that specifically binds the mannose that coats the surfaces of many tissues. Under tension, the mannose-bound FimH changes conformation, adopting a strong attachment state. This force-dependent attachment is known as a catch bond, which increases the reliability of cell attachment to the surface in strong flow environments but also leads to a "stick and roll" adhesion where cells slowly continuously move in the direction of the flow while remaining attached to the surface.

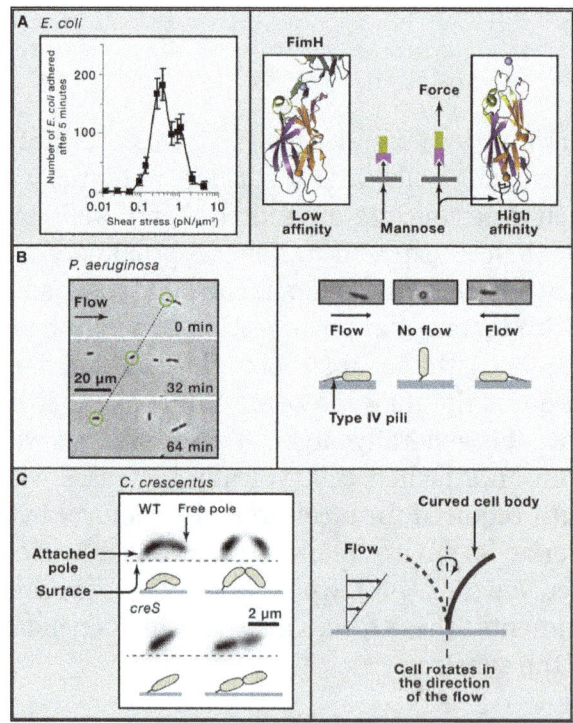

Bacteria leverage fluid flow at the scale of a single cell.

In the above figure, (A) E. coli use a catch bond mechanism to enhance attachment to surfaces in flow conditions. (left) Above a critical shear stress, cells are more likely to remain attached to a surface compared to cells that experience lower shear stress, as demonstrated by the peak in the number of adherent cells at finite shear stress. (right) The fimbrial capping protein FimH changes conformation when the fimbria is under tension, thereby increasing its affinity for surface-bound mannose. The mechanics of FimH are analogous to a finger trap toy, where extension enhances binding via twisting.

(B) (left) P. aeruginosa attaches to surfaces with polar pili and cells migrate via twitching motility in the direction opposite to the flow. (right) Flow reorients cells in the direction opposite to the flow; successive pili extensions and retractions promote upstream migration. (C) (left) C. crescentus reorients in the direction of the flow when growing on surfaces. (right) Hydrodynamic forces act on the curved C. crescentus cell body, attached from one pole, to orient its free pole toward the surface. Thus, new cells are born close to the surface and can better attach immediately following separation from the mother cell. A straight cell dividing in flow has its pole oriented away from the surface, thereby reducing the likelihood of attachment to the surface after separation, which then reduces the rate of surface colonization.

Catch bonds thus may be beneficial during gastrointestinal colonization, allowing cells to remain in a beneficial microenvironment by anchoring to the epithelium and modulating their shear resistance in response to flow conditions. Notably, uropathogenic strains of *E. coli* possess mutations in FimH that reduce the dependence of adhesion on shear stress, indicating that the benefit of a catch bond may be lost in a low frequency pulsatile flow environment. E. coli cells can further strengthen attachment by leveraging the mechanical deformation of type I fimbriae. These fibers extend under tension forces, so that the force applied to a single attached cell is distributed among multiple fimbriae, decreasing the load experienced by each fiber, and improving the ability of a cell to remain attached to a surface.

Shear stress also enhances the attachment properties of *Pseudomonas aeruginosa*, which exhibit longer residence times on surfaces when subjected to flow than under static conditions. However, unlike *E. coli*, P. aeruginosa employs a mechanism that is independent of surface chemistry. Indeed, when subjected to shear, adhesion of P. aeruginosa increases on both glass and elastomeric substrata. The flow-dependent increase in residence time of *P. aeruginosa* is diminished in mutants lacking polar type I and IV pili (*cupA1* and *pilC*), flagella (*flgK*), or the ability to synthesize certain EPS matrix components (*pelA*). While these observations do not entirely eliminate the possibility of attachment via catch bonds, the findings suggest an alternative mechanism of shear-dependent adhesion whereby multiple adhesive structures participate to increase surface attachment.

These two examples are not rare among bacteria, as shear-enhanced adhesion has been observed in a variety of other contexts. For example, other *E .coli* fimbral structures can form catch bonds with distinct ligands, and increased shear stress promotes the adhesion of *Staphylococcus epidermis* cell clusters to human fibrinogen-coated surfaces and of *S. aureus* to fibers of the mechanosensitive von Willebrand factor.

Upstream Migration

Some bacterial surface interaction mechanisms simultaneously enable surface attachment and locomotion. Type IV pili, for example, are cell-surface structures that can be rapidly polymerized and depolymerized, and cells use them to move over surfaces via successive pilus extension, tip attachment, and retraction, which altogether is termed twitching motility. *P. aeruginosa* and numerous other bacteria use twitching motility to explore surfaces prior to forming biofilms. Pili extension and retraction also promote intimate contact between single cells and the host during infection.

A striking architectural feature of type IV pili and other adhesive structures (*e.g.*, flagella and holdfasts) is their strict localization to the poles of many rod-shaped cells. In environments with flow, a cell attached to a surface via a polar appendage will experience forces that tend to align the cell body with the vicinal flow field. In *P. aeruginosa*, which attaches to surfaces using its polar type IV pili, this phenomenon produces the surprising flow-driven behavior of upstream motion. Under flow, pilus-attached *P. aeruginosa* cells align in the direction of fluid movement with the piliated pole facing upstream. By successively retracting and extending pili, such cells migrate upstream, against the direction of the flow, despite the force oriented opposite to them generated by shear stress. This behavior has also been observed in the plant pathogen *Xyllela* fastidiosa and E. coli harboring type I pili, and it may be a general feature of surface-attached species possessing motorized polar pili.

Cell Shape

Bacteria possess a wide variety of cell morphologies, and each species robustly maintains a characteristic shape by precisely coordinating complex cell wall synthesis machineries. The function of cell shape is likely to depend on the typical environment of each species, but the underpinnings of this cell shape-niche relationship remain unknown in the vast majority of cases. In some instances, however, there are hints about how cell shape may constitute an adaptation to specific environmental conditions. For example, the helical shape of *Helicobacter pylori* enhances swimming motility in hydrogels, a feature that aids cells in penetrating mucus layers during stomach infection. In contrast, the curved bacterium C. crescentus harnesses its shape and the mechanics of its hydrodynamic environment to enhance surface colonization. As mentioned above, C. crescentus cells attach to surfaces via a polar holdfast. In flow, surface-attached cells orient in the direction of the flow. Shear stress generates a torque on their curved cell bodies, which rotates them such that their unattached poles arc towards the substratum. Consequently, mother cells deposit newly born daughter cells onto the surface immediately downstream, which leads to the colonization of the downstream surface and the formation of a biofilm. Indeed, straight mutants of C. crescentus are less likely to have their progeny immediately attach to the surface following division, and such mutants are more frequently lost to the bulk flow. Thus C. crescentus may have evolved its curved shape to enhance surface colonization in environments with flow, indicating that bacterial morphology is potentially a result of adaptation to specific mechanical environments.

Touching Down

Bacteria adopt many phenotypes that can confer a fitness advantages when cells are associated with a substrate. Transitioning from a planktonic swimming state to surface attachment is presumably an expensive regulatory decision in terms of energetic and potential opportunity cost. In several notable cases, bacteria coordinate transitions between attachment to and detachment from surfaces, making specific use of mechanical cues transduced via cell-surface structures.

Swimming motility allows cells to explore the bulk of a fluid, but becomes largely unnecessary after surface attachment. Consequently, many flagellar systems possess a mechanism for disabling rotation in response to mechanical forces. In a low Reynolds number environment, an object moving very close to a boundary experiences a larger viscous force compared to that which it would experience far away from the surface. Relative to that of a planktonic cell, the rotating flagellum of a surface-attached cell experiences a significantly larger drag force, increasing the load on the flagellar motors. *E. coli* harnesses this hydrodynamic effect and subsequently alters flagellar rotation. More generally, many bacterial species exhibit behavioral changes upon inhibition of flagellar rotation. EPS secretion, for example, is strongly modulated in response to the load on flagella by B. subtilis and V. parahaemolyticus . Similarly, C. crescentus stimulates deployment of its holdfast using a flagellum-dependent mechanism when attaching to surfaces, strengthening adhesion when necessary. Recent work has suggested that bacteria also possess the means to translate surface contact into physiological changes in gene expression independently of flagellar function. However, the contributions of mechanics in all of the above examples remain to be determined quantitatively.

We note that other systems could potentially enable bacteria to mechanically sense surfaces. For example, the mechanosensitive channels protecting the integrity of the cell wall upon osmotic shock may also be sensitive to mechanical deformation of the cell wall and trigger surface-specific cellular responses, similar to those that occur among eukaryotes. Analogous to flagella-surface interactions, type IV pili are mechanically actuated cellular structures that could represent ideal mechanosensors, since their function is highly dependent on surface contact. Altogether, these mechanisms likely promote the colonization of surfaces and help to regulate the transition from the unicellular state to a multicellular lifestyle.

Mechanics at the Level of Multicellular Structures

In favorable environments, single surface-associated cells grow and divide or aggregate, thus initiating the formation of sessile, multicellular structures known as biofilms. When nucleated from a single founder cell, biofilms are often genetically homogeneous, although their finite sizes and spatial organizations generate distinct microenvironments for individual cells within them. For example, some cells occupy space near the substratum while others localize to the biofilm exterior, adjacent to the surrounding fluid. The availability of growth-limiting nutrients and other solutes often varies along sharp concentration gradients as a function of biofilm depth; as a result, a genetically homogeneous biofilm population can exhibit strongly heterogeneous phenotypes. In this manner and many others, physical and mechanical constraints affect bacterial behavior within biofilms. There is currently only limited quantitative and qualitative understanding of the effects of mechanics on multicellular bacterial development. Here, we highlight recent work exploring the influence of mechanics on biofilms, with emphasis on matrix rheology, fluid flow, and molecular transport.

Rheology

The biofilm matrix is a complex, heterogeneous gel-like material whose components vary from one bacterial species to another. The biophysical properties of the matrix have implications for the bacteria residing within biofilms, including their spatial organization and ecological interactions. For example, the matrix is dense and thus only poorly permeable to the surrounding fluid, ensuring that flow primarily occurs around biofilms. A consequence of this structural feature is that bacterial cells residing deep inside of biofilms receive only those nutrients capable of diffusing through the matrix. Additionally, cells residing in the outermost biofilm layers often rapidly deplete these nutrients, further denying access to cells in the interior. The resulting gradients in nutrient availability (and other solute concentrations) commonly generate physiological heterogeneity within biofilms, even those that are monoclonal, as cells within them adjust to their distinct local microenvironments.

Macroscopically, biofilms are examples of living soft matter composed of cells and EPS matrix. The matrix displays elastic, plastic, and viscous properties, allowing biofilms to distort under mechanical forces. The effective mechanical response to forces on biofilms resembles that of a viscoelastic fluid biofilm stiffness is affected by the chemistry of the environment and the growth state of the resident bacterial population. Biofilm expansion also generates mechanical stresses that can modify biofilm morphology and internal cellular organization. For example, in *Bacillus subtilis* biofilms growing at an air-liquid interface, compressive stresses generated by growth and expansion in a confined space leads to alteration of global biofilm morphology. In particular, *B. subtilis* biofilm sheets buckle to generate wrinkled pellicles at the surface of the fluid. At solid-liquid interfaces, stresses generate wrinkles at locations of the weakest matrix stiffness. Cell death locally reduces biofilm thickness, thus focusing mechanical stress or reducing local biofilm adhesion, leading to vertical buckling of the matrix. These morphological changes can generate a network of fluid-filled channels that increase permeability and enable nutrient transport by flow, thus improving nutrient availability to cells within the biofilm compared to purely diffusive transport. Similarly, the wrinkled morphology of *P. aeruginosa* biofilms increases their overall surface area, thereby improving oxygen uptake.

Flow

The flow around biofilms can also strongly affect their morphology, deforming their structures. Shear stress applied to attached bacteria can wash away secreted compounds, affecting biofilm density, limiting growth, and as a result, reducing the overall biofilm size. Following the transition from single attached cells to larger multicellular biofilm communities, mechanical effects can underpin the spread of bacteria to new locations. Large shear stresses generated under strong flow conditions may lead to biofilm breakage, dislodging bacteria from the community. Consequently, formerly biofilm-embedded bacterial cells are able to disperse to new territory and potentially colonize other environments that are more favorable.

One striking example of the effect of flow on biofilm architecture appears in irregular flow geometries. At corners and bends in curved channels, biofilms of some species develop as streamers – long, filamentous structures suspended in the fluid – nucleating at specific surface topological irregularities. Rusconi *et al.* showed that flow around corners promotes initiation of streamer formation. Biofilms nucleate at the downstream end of a bend and elongate in the direction of the flow, into the channel centerline. Numerical simulations suggest that the attachment point of a streamer co-localizes with flow moving fluid in the direction perpendicular to the main flow plane. The intensity of this secondary flow increases with sharper turns, which is consistent with the observation that streamers form more rapidly in such geometries.

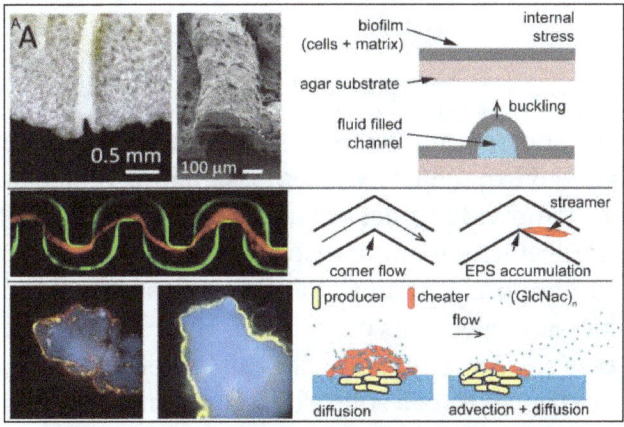

Effect of mechanics on multicellular communities.

In the above figure, (A) Photograph and scanning electron micrographs show fluidic channels within biofilms generated by buckling of the EPS matrix. Fluid evaporation through the biofilm generates flow within the channels leading to rapid advective transport of nutrients within the biofilm. In the absence of channels, nutrients only slowly diffuse into the biofilm. (B) Fluid flow promotes biofilm extrusions at channel bends that develop into fiber-like streamers extending into the channel centerline. These streamers form as channel bends induce localized flow patterns potentially favoring the accumulation of EPS. (C) The interplay between diffusive and advective transport of a nutrient shapes the interactions between "producer" enzyme-secreting cells (yellow) that digest a chitin substrate (blue) and mutant "cheater" cells that do not secrete the chitinase enzyme (red). Without flow, diffusion of liberated chitin oligomers (GlcNac) n permits "cheater" cells to exploit populations of chitinase-producing cells. In contrast, flow rapidly removes the soluble (GlcNac) n released from the chitin surface, denying access to "cheaters" and rendering secreted chitinase-production evolutionarily stable.

The formation of streamers therefore depends on the characteristics of the flow coupled with the topology of the surface. A critical feature of streamer growth, in contrast to surface-associated biofilms, is their spatial extension far into the bulk of the fluid. While wall-associated biofilms only modestly impair flow, streamers obstruct flow channels near the centerline and thus slow flow much more dramatically than do biofilms on the walls. This effect generates a positive feedback loop in which the decrement of flow favors accumulation of EPS and other suspended materials, until streamers completely clog the channel and catastrophically stop the flow. Because turns and bends are common within flow systems, such as an animal's vasculature or an industrial cooling system, we anticipate that streamer-induced clogging is a frequent impediment in many fluidic systems and networks. Consistent with this idea, streamers form in flow elements common to various environmental systems, including filters, soil, and stents. Many studies in this realm have focused on *P. aeruginosa* as a model organism; similarly S. aureus forms streamers that rapidly clog channels, and it does so in a manner that depends on the chemistry of the nucleating surface. Although only investigated recently, streamers are now predicted to be common when biofilms encounter bends and flow.

Transport

Bacteria survive, grow, and communicate by detecting and frequently importing compounds present in their environments. Nutrients, signaling molecules, and antimicrobials affect the behavior of individual bacteria and the development of their multicellular communities. These molecules generally reach cells by diffusion, but flow can dramatically modify the spatiotemporal distribution of such compounds and the effective concentrations that surface-attached cells experience. In practice, the contribution of flow to the transport process can be estimated with the Péclet number, which measures the relative contribution to transport of a solute by advection (mediated by flow) compared to the contribution by diffusion ($Pe = UL/D$, where U is a typical value of flow speed, L is the length scale of the system, e.g. of the biofilm, and D is the diffusivity of the solute). For example, a Péclet number much larger than unity signifies that solutes mainly move with the flow. Conversely, the transport of a solute is dominated by diffusion when the flow is slow enough to yield a Péclet number much smaller than one.

In many instances, nutrients are not derived from the bulk fluid surrounding a biofilm, but rather from the substrata to which the cells are attached. In such cases, biofilm-dwelling cells often digest the substrata on which they reside by secreting extracellular enzymes, liberating nutrients that can freely diffuse away. This scenario generates a public goods conundrum: cells that secrete digestive enzymes can be exploited by other species that fail to produce the enzyme but nonetheless benefit from its production. Recent work has shown that a flow environment can help to overcome this evolutionary dilemma. Outside of its human host, the pathogen *Vibrio cholerae* grows on solid particles of the biopolymer chitin, secreting chitinase enzymes that digest chitin and liberate the soluble product N-acetylglucosamine (GlcNac) and oligomers, which are excellent nutrients for growth. In the absence of flow, a "cheater" (a mutant that does not produce chitinase) can scavenge diffusing nutrients released by the producers' chitinase activity and outcompete the producer because the cheater does not pay the cost of producing chitinase. In contrast, flow disperses nutrients generated by producers away from the chitin granules, limiting growth exclusively to chitinase-producers and their offspring who are residing on the chitin surface; cheater mutants are essentially starved out of the system.

Similarly, flow transports signaling molecules and other secreted compounds that mediate social interactions within and between bacterial populations. An important example is how flow affects quorum sensing, an intra- and inter-species bacterial communication system that is used to synchronize collective behaviors. Quorum sensing relies on the production, release, and detection of extracellular signal molecules called autoinducers. In well-mixed environments, cells assess cell density by measuring the local autoinducer concentration to initiate group behaviors. By contrast, in a heterogeneous environment, for example in a sessile population in flow, advection and diffusion affect local autoinducer concentrations. Bacteria may, in fact, use quorum-sensing systems to detect the flow in their surroundings, thus probing their vicinal surroundings for their growth potential or for other cues.

Connecting Scales

Fully understanding bacterial behavior requires efforts to address the common and intimate association of cells with surfaces. We must consider individual cells, their morphology, and their responses to environmental cues and mechanical forces such as those described above: surface adhesion in the presence of flow, solute transport, and biofilm rheological properties. Cell division and matrix secretion combine to promote biofilm formation on the scale of many hundreds to millions of cells for which flow, transport, and rheology feed back onto the population dynamics within biofilm populations. Clarifying the consequences of mechanics for bacterial cells in isolation and as members of collectives is therefore central not only to understanding the transitions between individual and multicellular bacterial behavior but also to bacterial evolution in the broadest sense.

The phenomena, which largely pertain to the behavior of single cells, influence transitions from surface occupation by individual bacteria to the formation of large multicellular communities. An adaptation that first appears to benefit only individual cells could influence the fitness of its descendants many generations later. A prime example of such adaptation is the increased ability of C. crescentus to colonize surfaces in flow as described above. Cell curvature increases the rate of surface attachment by daughter cells, thereby increasing the rate of surface colonization and ultimately enabling capture of increased niche space in a three-dimensional biofilm matrix that may extend many cell lengths away from the substratum. Populations of curved cells thus form robust biofilms more rapidly than straight cells, such that curvature may be viewed as adaptive not only at the scale of a single colonizing bacterium but at the scale of clonal populations descended from surface-associated founder cells. Similarly, the upstream motility of single P. aeruginosa cells provides a group-level advantage when competing with planktonic cells and other species during colonization of fluidic networks. Phenotypes that evolved in response to mechanical forces experienced by single surface-attached cells could thus contribute to pathogenic infection by a large bacterial population.

When forming biofilms, bacteria implement individual gene expression programs that ultimately contribute to the properties of the entire community. As individual cells respond to or modify the chemical and mechanical features of their environment, they express phenotypes and generate cell group configurations that contribute to collective survival. For instance, impaired flagellar rotation upon surface contact has been proposed as a signal that induces biofilm formation, as motility and EPS secretion are negatively correlated. Thus, surface contact generates a mechanical cue triggering an intracellular signaling cascade that leads to many phenotypes, which culminate in stable adhesion to the substratum. Upon growth and division, bacteria modify their mechanical environment by secreting EPS that helps maintain the spatial coherence of clonal lineages within nascent biofilms and discourages invasion by planktonic cells. These and other recent studies together suggest that biofilms often emerge from the behavior of individual cells that are locally cooperative within their strain or clonal lineage and globally competitive with other strains and species with which they may be growing.

Subsequent to the formation of multicellular communities, biofilm structure and mechanical properties feed back upon the forces and solute concentrations experienced by biofilm-dwelling bacteria to further influence their individual and collective behaviors. For example, because advection is negligible within the matrix, solute transport occurs primarily through diffusion, which in turn, leads to heterogeneous concentration gradients of soluble compounds secreted or absorbed by individual bacteria. These solute distributions further influence the competitive dynamics within and among competing cell lineages inside biofilms. Biofilm structure and diffusive properties likewise influence the accumulation of autoinducers involved in the quorum-sensing-regulated activation of group-wide phenotypes such as virulence in P. aeruginosa. In V. cholerae, by contrast, quorum sensing feeds back into the structure and rheology of the biofilm to repress EPS at high cell density, thus initiating dispersal of the pathogen.

Do single cells actively sense mechanical forces? What is the relevance of these features in realistic ecological contexts? How do biofilms form in complex and diverse environments such as the digestive tract? Do organ mechanics affect bacterial population development? These and many other questions are natural extensions of past and current experimental explorations of bacterial behavior. Resolving the connections between mechanics and biology in the bacterial world will require integrative approaches that combine genetics, biochemistry, chemistry, evolutionary biology, physics, and engineering principles.

Surface-specific mechanics are ubiquitous in the bacterial world, and the examples above highlight the essential role that they play in many elements of bacterial processes. The insights gained from a mechanics-informed view of bacteria will, in turn, improve our ability to control microbes in settings where they can be helpful or harmful to humans. For instance, a better understanding of how mechanics contribute to regulating virulence may provide alternative approaches to fight infections and help overcome the rise of antibiotic resistance. More generally, bacterial mechanics represent an exciting research direction for biologists aiming to understand bacterial physiology in realistic environments and for engineers and physicists aiming to develop new tools and models to interface with microbiology and develop a fully interdisciplinary understanding of bacterial behavior.

References

- Bacteriology, cell-biology, biology-and-genetics, science-and-technology: encyclopedia.com, Retrieved 16 June, 2019

- James franklin crow; william f. Dove (2000). Perspectives on genetics: anecdotal, historical, and critical commentaries, 1987-1998. Univ of wisconsin press. P. 384. Isbn 978-0-299-16604-5

- Bacteria-cell-meaning-and-structure-with-diagram, bacteria: biologydiscussion.com, Retrieved 24 March, 2019

- Gitai z (2005). "the new bacterial cell biology: moving parts and subcellular architecture". Cell. 120 (5): 577–86. Doi:10.1016/j.cell.2005.02.026. Pmid 15766522

- Morphology-and-classification-of-bacteria: researchgate.net, Retrieved 13 July, 2019

- Miller a, korem m, almog r, galboiz y (june 2005). "vitamin b12, demyelination, remyelination and repair in multiple sclerosis". Journal of the neurological sciences. 233 (1–2): 93–7

- Bacteria-definition-morphology-classification-and-reproduction-microbiology, bacteria: biologydiscussion.com, Retrieved 12 April, 2019

- Choi cq (17 march 2013). "microbes thrive in deepest spot on earth". Livescience. Archived from the original on 2 april 2013. Retrieved 17 march 2013

- Factor-affecting-bacterial-growth: onlinebiologynotes.com, Retrieved 21 March, 2019

5

Industrial Applications of Bacteria

Bacteria are used in various industries for diverse purposes such as food manufacturing production of probiotics, antibiotics, drugs, vaccines, insecticides, enzymes, fuels and solvents. The chapter closely examines these industrial applications of bacteria to provide a thorough understanding of the subject.

MEDICAL USES FOR BACTERIA

There are ten times more of them in and on our bodies than human cells, so it is perhaps surprising that the 10,000 species of bacteria in the body have not previously been harnessed to any great extent to improve health. But as genetic science advances, research is pointing towards bacteria as a useful diagnostic tool in the detection of cancer, diabetes, Crohn's disease and periodontitis.

Researchers have genetically modified Escherichia coli bacteria to emit light in the presence of tumours or glucose. In the first experiment, E coli that grew on particular types of tumour while ignoring healthy tissue were harvested from a readily available probiotic. The bacteria were modified to produce an easily detectable enzyme and given to mice with liver tumours. The mouse urine changed colour or gave off light depending on which additional substances were added to the signalling system.

In a second experiment, also recently published in Science Translational Medicine, E coli were genetically engineered to change colour in the presence of sugar. They then changed colour when added to urine that contained glucose.

Separate research has shown that the metabolism of bacteria living in the mouth changes significantly when the person has certain conditions. Researchers at the University of Texas used a supercomputer to map the metabolic pathways of 60 different species of bacteria. They analysed more than 160,000 genes to create a 'subway map' using different coloured lines to denote pathways that were unchanged between disease and health, and those that were upregulated during either health or disease.

Periodontitis, one of the most prevalent diseases in the world, was chosen to study because broadly the same bacteria are present in the mouth in both healthy and disease states. But oral bacteria also change behaviour in diseases such as diabetes and Crohn's disease. This research could be used to develop diagnostic biomarkers for certain conditions. If the bacteria could be modified back to their 'healthy' state this technology could even be used as a treatment.

But genetic modification outside of the lab remains controversial, and use of GM organisms has been limited by fears of the impact of GM material on the environment. Scientists have attempted to solve the potential problem of GM bacteria escaping into the wild in a variety of ways, including the creation of bacteria that depend on nutrients they can't make themselves.

But these bacteria might survive in the wild by receiving those nutrients from natural organisms. One potential solution would be to create genetically engineered bacteria that are dependent on nutrients not found in nature.

APPLICATIONS OF BACTERIA IN INDUSTRY AND BIOTECHNOLOGY

Bacteria are used in industry in a number of ways that generally exploit their natural metabolic capabilities. They are used in manufacture of foods and production of antibiotics, probiotics, drugs, vaccines, starter cultures, insecticides, enzymes, fuels and solvents. In addition, with genetic engineering technology, bacteria can be programmed to make various substances used in food science, agriculture and medicine. The genetic systems of bacteria are the foundation of the biotechnology industry.

In the foods industry, lactic acid bacteria such as Lactobacillus, Lactococcus and Streptococcus are used in the manufacture of dairy products such as cheeses, including cottage cheese and cream cheese, cultured butter, sour cream, buttermilk, yogurt and kefir. Lactic acid bacteria and acetic acid bacteria are used in pickling processes such as olives, cucumber pickles and sauerkraut. Bacterial fermentations are used in processing of teas, coffee, cocoa, soy sauce, sausages and an amazing variety of foods in our everyday lives.

In the pharmaceutical industry, bacteria are used to produce antibiotics, vaccines, and medically-useful enzymes. Most antibiotics are made by bacteria that live in soil. Actinomycetes such as Streptomyces produce tetracyclines, erythromycin, streptomycin, rifamycin and ivermectin. Bacillus and Paenibacillus species produce bacitracin and polymyxin. Bacterial products are used in the manufacture of vaccines for immunization against infectious disease. Vaccines against diphtheria, whooping cough, tetanus, typhoid fever and cholera are made from components of the bacteria that cause the respective diseases. It is significant to note here that the use of antibiotics against infectious disease and the widespread practice of vaccination (immunization) against infectious disease are two twentieth-century developments that have drastically increased the quality of life and the average life expectancy of individuals in developed countries.

Biotechnology

The biotechnology industry uses bacterial cells for the production of biological substances that are useful to human existence, including fuels, foods, medicines, hormones, enzymes, proteins, and nucleic acids. The possibilities of biotechnology are endless considering the gene reservoirs and genetic capabilities within the bacteria. Pasteur said it best, "Never underestimate the power of the microbe."

Biotechnology has produced human hormones such as insulin, enzymes such as streptokinase, and human proteins such as interferon and tumor necrosis factor. These products are used for the treatment of a various medical conditions and diseases including diabetes, heart attack, tuberculosis, AIDS and SLE. Botulinum toxin and BT insecticide are bacterial products used in medicine and pest control, respectively

One biotechnological application of bacteria involves the genetic construction of super strains of organisms to perform particular metabolic tasks in the environment. For example, bacteria which have been engineered genetically to degrade petroleum products are used in cleanup of oil spills and other bioremediation efforts.

Another area of biotechnology involves improvement of the qualities of plants through genetic engineering. Genes can be introduced into plants by a bacterium Agrobacterium tumefaciens. Using A. tumefaciens, plants have been genetically engineered so that they are resistant to certain pests, herbicides, and diseases.

Finally, it should not be overlooked that industrial, pharmaceutical and food microbiology are applications of biotechnology. Archaea and bacteria are involved in production of biofuels. Bacteria are the main producers of clinically useful antibiotics; they are a source of vaccines against once dreaded diseases; they are probiotics that enhance our health; and they are primary participants in the fermentations of dairy products and many other foods.

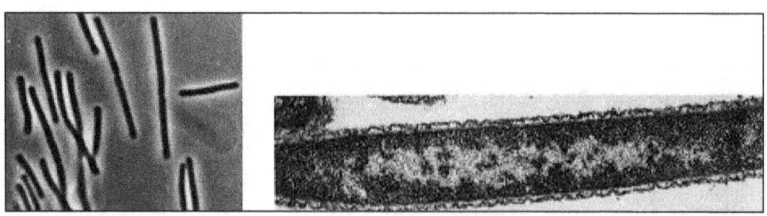

Thermus aquaticus, the thermophilic bacterium that is the source of taq polymerase. L wet mount; R electron micrograph. T.D. Brock. Life at High Temperatures. The

polymerase chain reaction (PCR), a mainstay of the biotechnology industry because it allows duplication of genes starting with a single molecule of DNA, is based on the use of the DNA polymerase enzyme derived from Thermus aquaticus.

Bacteriotherapy

Bacteriotherapy is the purposeful use of bacteria or their products in treating an illness. Forms of bacteriotherapy include the use of probiotics, microorganisms that provide health benefits when consumed; fecal matter transplants (FMT)/intestinal microbiota transplant (IMT), the transfer of gut microorganisms from the fecal matter of healthy donors to recipient patients to restore microbiota; or synbiotics which combine prebiotics, indigestible ingredients that promote growth of beneficial microorganisms, and probiotics. Through these methods, the gut microbiota, the community of 300-500 microorganism species that live in the digestive tract of animals aiding in digestion, energy storage, immune function and protection against pathogens, can be recolonized with favorable bacteria, which in turn has therapeutic effects.

FMT is being used as a new and effective treatment for C. diff infections, a gastrointestinal disease in which Clostridium difficile colonizes the gut of an organism disrupting microbial balance and causing diarrhea that can potentially be deadly. Bacteriotherapy has also begun to be used in the treatment of mental disorders such as depression, anxiety, and Autism Spectrum Disorder. Recolonization of gut flora can be used effectively in the treatment of mental disorders because of the existence of the gut-brain axis, the bidirectional route of communication between the brain and the gut, specifically the gut microbiota.

Fecal Matter Transplant (FMT)

Fecal Matter Transplant (FMT) was first documented in humans in 1958. The FDA considers FMT a suitable treatment for select patients with C. diff, specifically when standard treatment has failed. It shows a 90% success rate in clinical trials for recurrent C. diff infections. For other illness, it is considered an experimental treatment and should only be done within a research program.

The process of FMT involves injecting a liquid suspension of healthy stool into the gastrointestinal tract of a patient. FMT does not require immunological matching or suppression (unlike typical organ transplants). FMT can be performed through nasogastric intubation, nasojejunal intubation, nasoduodenal intubation, upper tract endoscopy, retention enema, gentle rectal enema, or colonoscopy. Research is currently being done to see if FMT can be encapsulated and taken orally as a pill.

FMT is a novel treatment, with few complications known thus far. Minor side effects have been reported as mild diarrhea, cramping, abdominal pain, changes in bowel movements, upper gastrointestinal hemorrhage, IBS symptoms (infectious or not), constipation, and irritable colon. There is little known about the possible long-term risk of transmitting an autoimmune disease. Protocols vary with regard to quantity of stool being transplanted and method of infusion.

Fresh unfrozen stool samples are more commonly used than frozen samples. The transplant of unfrozen sample is preferably completed within 6 hours. Resolution and relapse rates also differ based on the diluents used to make FMT solutions (water, saline, yogurt, milk or saline with psyllium). Resolution rates increased with increased volume and relapse rates increased with decreased mass of FMT.

Further research on FMT is required to directly compare routes of administration, optimal protocol for infusions, and ideal amounts of fecal matter required. Researchers suggest using a large sample size in order to yield statistically significant results.

FMT Donation

Donors must have refrained from antibiotic usage for as little as 2 months or up to 6 months prior to donating stool. Additionally, donors must not have any history of gastrointestinal disease. Blood tests commonly screen donors for hepatitis A, B and C, HIV and syphilis. Stool tests may include CD toxin, ova, and parasites. 1 donor provides feces for more than 1 patient. Fresh donations should be provided on the day of treatment. Donors related to recipients typically show higher resolution rates (93%) compared to unrelated donors (84%).

Medical Uses

C. Diff

C. diff infections typically result from the use of broad spectrum antibiotics that alter the microbiota balance, allowing C. diff to colonize. Typical treatment of C. diff with antibiotics, can further disrupt the microbiome of the gut often leading to a cyclical recurrence of C. diff with 35% of patients experiencing recurrence, additionally antibiotic resistance is a growing problem.

Replacing typical antibiotic treatments with novel FMT treatment restores healthy microbiota and resolves symptoms. FMT restores a healthy balance of bacteria within a gut previously disrupted by C. diff colonization and antibiotic usage. FMT colonizes the gut with microbiota that suppress C. diff, rebuilds a stable microbiome, and restores function. The microbiota of treated patients typically resembles that of the donor after transplantation.

In a systematic review of the use of FMT to treat C. diff infections (mostly C. diff associated diarrhea), 536 patients age 4-77 were reviewed, with elderly patients predominating. Most patients had previously received antibiotics before having repeated relapses. 87% of patient's diarrhea resolved after first FMT treatment. Diarrhea resolution rates differed based on injection: 81% resolution when injected in the stomach, 86% when injected into the duodenum/jejunum (the first two parts of the small intestine), 93% success upon transfer by colonoscopy into the cecum (pouch at the junction of the small and large intestine)/ascending colon, 84% when inserted into the distal colon. Upon resolution of diarrhea, C. diff toxin tests were found negative.

Another systematic review of 317 patients age 2-95, (average of 53 years) showed resolution of C. diff 92% of the time after treatment by FMT. Stool transplants were typically greater than or equal to 200 mL. 89% showed resolution after 1 treatment. Infusion by gastroscopic/nasojejunal tube showed the lowest resolution rates at 76%.

Administration Methods for C. Diff

Administration via colonoscopy showed highest rates of successful clearance of C. diff associated diarrhea, indicating that the direct deliverance of healthy bacteria to the site where the majority of C. diff is established is the most successful therapeutic avenue. Additional benefits of colonoscopy are recolonization with favorable bacteria, bowel cleaning to rid residual C. diff spores, injection of a larger volume of stool sample than other methods and it allows visualization of the colon to rule out other disease.

Upper endoscopy and nasogastric tubes are commonly used to avoid performing an endoscopy through an inflamed colon where there is a small risk of perforation. Additionally, endoscopy is a slow procedure; however, the disadvantages of the upper endoscopy and nasogastric tube methods are that the sample cannot be inserted directly into the C. diff affected site in the colon and the sample may be degraded before reaching the colon—accounting for its lower resolution rate. Enema is a less expensive and less invasive option.

Administration routes are typically decided on a case-by-case basis and account for some differences in resolution and relapse rates.

Mental Illness: Depression/Anxiety

There are high levels of comorbidity between some mental disorders and gastrointestinal disturbances. This provides support for the existence gut-brain-axis, in which microbiota of the gut can influence brain development, function, and behavior, and emphasizes its role in mental illness, making FMT a plausible therapeutic avenue for some mental illness.

Microbiota modulate the hypothalamic-pituitary-adrenal-axis (HPA axis), which controls reactions to stress and regulates digestion, immune system, mood, and emotions. Additionally, microbiota can directly impact the central nervous system (CNS), as studies have shown that bacteria in the gut can activate stress response through the vagus nerve, a cranial nerve responsible for interactions with the digestive tract. Evidence suggests that while stress can impact the composition of the microbiome, the microbiome also has an impact on stress response and behavior.

Research on FMT has shown that upon transfer of fecal matter from a donor to a germ-free recipient, animals that have no microorganisms living in them, the recipient begins to mimic the phenotype, observable characteristics, of the donor. In experiments with mice, an obese donor led the recipient mouse to adopt an obese phenotype, while an underweight donor led to the adoption of an underweight phenotype in the recipient. It is thought that this mechanism explains why FMT is an effective treatment for depression and anxiety, as well as obesity.

The microbiome of depressed people has been found to show decreased richness and diversity. Specifically, lactobacillus and bifidobacterial have been identified as having roles in modulating depression and anxiety behaviors. When fecal matter is transferred from depressed mice to microbiota depleted mice, behavioral exams show anhedonia, a symptom of depression; studies have also found that in a microbiome transfer from a stressed animal to a control, the control recipient also exhibits anxious behaviors proving that some depression and anxiety phenotypes are dependent on the gut microbiome, and therefore transferrable. The identification of the microbiome as having a causal role in depression and anxiety, as well as the ability to transfer depressive or anxiety symptoms from a depressed donor to a recipient makes the reverse, FMT from a healthy donor to a depressed or anxious patient, a valid experimental treatment for depression.

Mental Illness: Autism Spectrum Disorder (ASD)

The gut microbiome has been implicated in Autism Spectrum Disorder (ASD) due to its high comorbidity with gastrointestinal problems correlating with severity of ASD. Mouse models of ASD show a link between abnormal metabolites in the gut and behavior. A clinical trial of 18 ASD children undergoing 2-week antibiotic treatment, bowel cleanse, followed by extended FMT showed an 80% reduction of gastrointestinal symptoms. Behavioral ASD symptoms also showed significant improvement that persisted up to 8 weeks after treatment ended. These changes are attributed to colonization of donor microbiota and beneficial changes in the gut environment.

Probiotics

Probiotics are living bacteria or fungi that confer health benefits. They have 3 mechanisms of therapeutic effect: antimicrobial effects, strengthening lining of the intestines, and immune modulation. These mechanisms help alter and diversify gut flora to benefit overall health. The antimicrobial effect helps prevent the growth of bacteria that cause illness. Probiotics also help strengthen tight junctions, multiprotein complexes lining the intestines (as well as other organs and regions of the body) to prevent passage of materials. Leaky gut is the term used when tight junctions of the intestines are disrupted and bacteria can exit the intestines. Leaky gut is implicated in many gastrointestinal disorders. Probiotics stimulate production of a protein that maintains the strength of the intestinal barrier and have beneficial effects on local and systemic immune system responses.

Medical Uses

Gastrointestinal Disorders

Probiotics have been used in treatment or prevention of C. diff, irritable bowel disease, irritable bowel syndrome, prevention of radiation or chemotherapy induced sequelae, necrotizing enterocolitis, hepatic encephalopathy, and atopic dermatitis. Success in treatment depends on whether single or mixed strains are administered, dose, and specific bacterial species. Lactobacillus and Bifidobacterium are the most commonly used probiotic strains.

Use of probiotics in psychological states and autism is currently being studied and has shown that probiotics may influence psychological states. Probiotics have been used to transfer neurochemicals such as GABA. There is evidence that disruption of the microbiome may promote overproduction of clostridium tetani, a neurotoxin producing bacteria that may contribute to symptoms of autism. One case study on a 12-year-old boy with ASD, severe cognitive disability, and celiac disease who received probiotic treatment for celiac showed an unexpected improvement in autistic core symptoms that persisted 4 months after treatment. Administration of an Autism Diagnostic Observation Schedule (ADOS) showed a 3-point decrease in score (i.e. an improvement in core autistic symptoms) in the social affect domain. ADOS scores are typically consistent measures of autism severity, and change is unlikely. Microbiota reports consistently show significant differences in the gut of autistic patients compared to non-autistic. Ongoing research investigating probiotic use for ASD treatment is required to determine the efficacy of probiotics in reducing symptoms of ASD, and the correlation between probiotics, gut microbiota, and ASD symptoms.

In studies with mice, probiotic treatment reduced anxiety and depressive behaviors, reversed the impact of maternal separation on depressive behaviors, reversed inflammatory induced and parasite induced anxiety behaviors. This evidence suggests that probiotic treatment has antidepressant and anxiolytic effects.

Synbiotics

Synbiotics contain both prebiotics and probiotics. Combining prebiotics with probiotics improves survival and activity of probiotic bacterial species. In synbiotics, prebiotics and probiotics work synergistically to provide a combined benefit beyond what either could confer independently. Synbiotics have shown positive effects on obesity, diabetes, non-alcoholic fatty liver disease, necrotizing enterocolitis in very low birth weight infants, and hepatic encephalopathy. Synbiotics can be used both as preventative measures and therapeutic treatments.

BACTERIAL COMPUTING

The capability to establish adaptive relationships with the environment is an essential characteristic of living cells. Both bacterial computing and bacterial intelligence are two general traits manifested along adaptive behaviors that respond to surrounding environmental conditions. These two traits have generated a variety of theoretical and applied approaches. Since the different systems of bacterial signaling and the different ways of genetic change are better known and more carefully explored, the whole adaptive possibilities of bacteria may be studied under new angles. For instance, there appear instances of molecular "learning" along the mechanisms of evolution. More in concrete, and looking specifically at the time dimension, the bacterial mechanisms of learning and evolution appear as two different and related mechanisms for adaptation to the environment; in somatic time the former and in evolutionary time the latter.

Bacterial computing is an applied field recently launched, as well as the theoretical approaches to prokaryotic or bacterial intelligence, are derived from the adaptive response of living cells to existing environmental conditions. From a practical standpoint, we could define bacterial computing as the possibility of using bacteria for solving problems that today are solved by computers. If a bacterium could perform the work of a computer, this would allow us to build millions of computers which be replicated every 30 min, and that they would be confined within a Petri dish. According to Amos natural computing paradigms inspired by biological processes (e.g., artificial neural networks, genetic algorithms, ant colony algorithms, etc.) have proved to be very effective. However, all these "forms of computing" occurs *in silico*, and therefore within a computer. At present and in agreement with Amos, the challenge is the possibility to use biological substrates and biological processes to encode, store and manipulate information. For instance, to build a simple computing device, using bacteria rather than silicon. Since the seminal work of the feasibility of using biological substrates for computing has been well-established: Levskaya et al. has shown that the living cell could be considered as a programmable computational device, Baumgardner et al. using DNA segments and Hin/hixC recombination system successfully programmed *E. coli* with a genetic circuit that enables bacteria to solve a classical problem in artificial intelligence, the Hamiltonian problem; or the theoretical model where bacteria are used to solve the "burnt pancake problem.

In a theoretical realm, bacterial computing could be an emergent phenomenon consequence of learning and evolution. Bacterial learning and evolution are but two different and related mechanisms for adaptation to the environment, in somatic time the former and in evolutionary time the latter.

During recent years, some experiments have shown that bacteria can learn the ability to anticipate changes in their immediate environment. For instance, Tagkopoulos et al. found how *E. coli* colonies can develop the ability to associate higher temperatures with a lack of oxygen, and how bacteria have naturally "learned" to get ready for a serving of maltose after a lactose appetizer. According to Tagkopoulos et al., homeostasis explains microbial responses to environmental stimuli—by means of intracellular networks, microbes could exhibit predictive behavior in a fashion similar to metazoan nervous systems. Even more, bacteria are able to explore the environment within which they grow by utilizing the motility of their flagellar system and deploying a sophisticated "chemotactic" navigation system that samples the environmental conditions surrounding the cell and systematically guides *away* from the unfavorable conditions and *toward* the favorable ones.

Bacteria can also evolve some true learning behaviors to respond optimally to their environment (Bacterial evolution). At present, most methods in evolutionary computation are inspired in the fundamental principles of neo-Darwinism or population genetics theory, considering as main sources of variability chromosome crossover (or recombination) and mutation. However, bacteria exhibit several other genetic mechanisms as sources of variability, i.e., mechanisms such as transformation, conjugation, and transduction. In Bacterial conjugation we discuss the applicability of conjugation, a genetic mechanism exhibited by bacterial populations, and we simulate the evolutionary process along this mechanism. The efficiency of bacterial conjugation is illustrated designing, by means of a genetic algorithm based on this very mechanism, an AM radio receiver. In Bacterial transduction we continue the study of bacterial evolution, in this case by modeling and simulating the whole bacterial transduction mechanism.

Modeling Proteins as McCulloch-Pitts Neurons

Adaptive behavior in bacteria depends on organized networks of proteins governing molecular processes within the cellular system. Bacteria are able to explore the environment within which they develop by utilizing the motility of their flagellar system as well as a biochemical navigation system that samples the environmental conditions surrounding the cell. The McCulloch-Pitts artificial neuron is a mathematical model of a biological neuron. A neuron has a set of inputs $I_1, I_2, ..., I_i$, a set of weight values associated with each input line $W_1, W_2, ..., W_i$, one output O_i and a linear threshold function $f(net_i)$. Every time a neuron receives a set of input signals, performs the weighted sum (with the weights associated with each line) obtaining a net_i value, finally deciding its state or output with a threshold function:

$$net_i = \sum_i w_i I_i$$

$$O_i = \begin{cases} 1 & net_i \geq \theta \\ 0 & net_i \geq \theta \end{cases}$$

where θ is a threshold value.

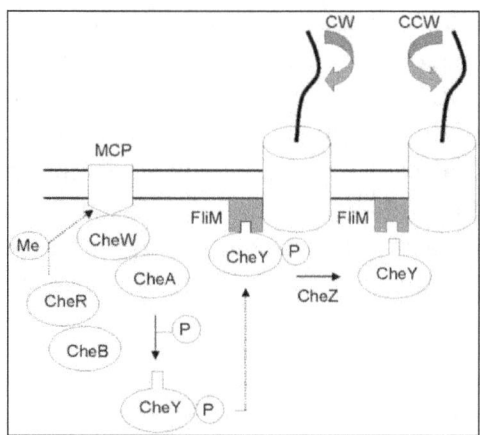

Figure: Che protein network mediating bacterial taxis. External factors are identified by membrane receptors (MCP) which reactivity depends on methylation levels controlled by CheB and CheR cytoplasmic proteins. In the network CheW is the bridge between the receptors and protein kinase CheA which can donates a phosphate group to CheY. The binding of CheY–P to the switch complex protein FliM induces a clockwise flagellar rotation (CW) whereas the dephosphorylation of CheY (via CheZ) restores counter-clockwise flagellar rotation.

The hardware model of the McCulloch-Pitts output artificial neuron assumes some plausible analogies between neurons and proteins. For instance, the connection between neurons, thus synapses as well as the neuron input and output would have their equivalent in the proteins on the concepts of bond strength, external factors (e.g., heat, ions, chemical substances, etc.) and conformational states related to catalysis and binding, respectively. In this analogy, it is also assumed that activation function represents cooperativity in proteins.

The hardware model of the proteins as a McCulloch-Pitts artificial neuron is based on a 741 inverting operational amplifier. The operational amplifier simulates the bond strength, external factors, conformational states and cooperativity by means of a potentiometer $(R_1, R_2, ..., R_n)$, voltage $(V_1, V_2, ..., V_n)$, voltage (V_0) and its saturation curve. In the circuit $R_1, R_2, ..., R_n$ are n variable resistors modeling the weights associated to connections among the n membrane receptors—playing the role of the input neurons—and the McCulloch-Pitts output neuron. Thus, variable resistors simulate the degree of influence of n inputs or external factors acting as *repellents* assuming one input per membrane receptor. The resistor R_f simulates the feedback in the output neuron being V_0 the output voltage. Assuming the presence in the medium or environment of n inputs or external factors which presence is modeled as V_i ($i = 1, ..., n$) input voltages with R_i ($i = 1, ..., n$) weights or variable resistors. According this model it follows that:

$$V_0 = -R_f \sum_{i=1}^{n} \frac{V_i}{R_i}$$

Once the input is applied, the value of the output is given by the LED diode state powered by the V_0 voltage. Even though LED diode intensity changes according to the output voltage in our experiments we only considered two states of the LED diode. When LED diode state is switch-off state then it means an output equal to 0 simulating that CheY and FliM remain separated and by consequence the flagellar rotation is CCW, indicating the bacterium swimming behavior. Otherwise, when the LED diode is switched-on then it means an output equal to 1, simulating that CheY binds to FliM, and as a result the flagellar rotation is CW, indicating a bacterium tumbling behavior. The threshold between states simulates the critical level of phosphorylation in which CheY binds to FliM resulting in the transition from 0 (swimming behavior) to 1 (tumbling behavior).

In the practical implementation of the model, the amplification factor A of the operational amplifier:

$$A = -R_f \sum_{i=1}^{n} \frac{1}{R_i}$$

simulates the catalytic effect of enzymes, e.g., the protein kinase CheA which donates a phosphate group to CheY resulting a phosphorylated CheY. As a consequence, if we establish a similarity between the R_f resistor and the Michaelis–Menten K_m constant then the resistor R_f would be simulating the affinity between the enzyme and substrate. Note that in our hardware model the inhibitory connection that is usually included in the hardware implementation of artificial neural networks has been removed.

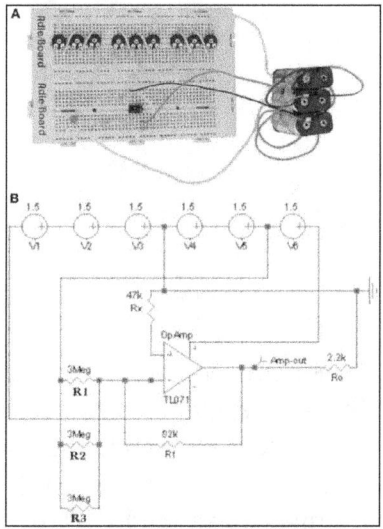

(A) Hardware and (B) software operational amplifier circuits with three repellent inputs. A voltage of +3V is applied to the potentiometers while the amplifier is powered with +9V (+4.5 and −4.5 V). In the hardware circuit (A) a LED diode connected to the amplifier output detects the intensity of the device signal. In the software circuit (B) a virtual voltage probe replaces to the LED diode.

Modeling Proteins as Networks of Processing Elements

In general, the molecular systems involved in bacterial signaling (and in *M. tuberculosis*) are extremely diverse, ranging from very simple transcription regulators (single proteins comprising just two domains) to the multi-component, multi-pathway signaling cascades that regulate crucial stages of the cell cycle, such as sporulation, biofilm formation, dormancy, pathogenesis, etc. A basic taxonomy of bacterial signaling is shown in figure. The first level of complexity corresponds to the simplest regulators, the "one-component systems." Actually, most cellular proteins that participate in cellular adaptation to the changing environment, in a general sense, could be included as participating within this elementary category. Following the complexity scale is the "two-component systems," which include histidine protein-kinase receptors and an independent response regulator; they have been considered as the central signaling paradigm of the prokaryotic organisms, since a number of intercellular and interspecies communication processes are served by these systems. A further category (conceptual consistency) of "three-component systems" is applied to those two-component systems that incorporate an extra non-kinase receptor to activate the protein-kinase.

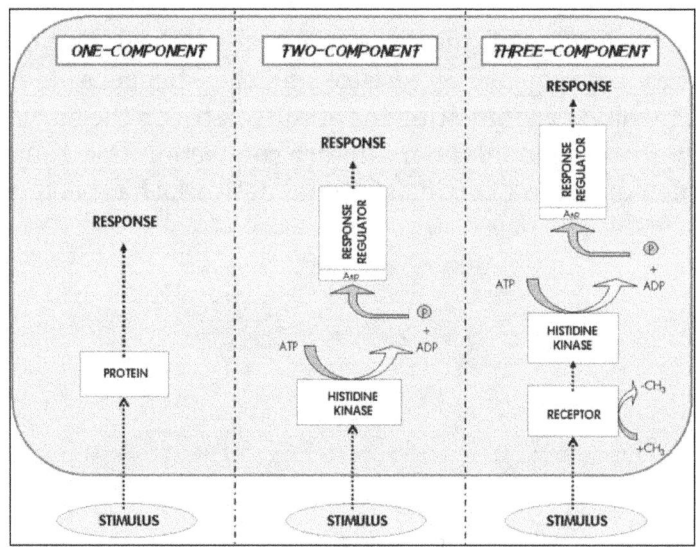

Figure: The one-two-three Component Systems. These three systems are the characteristic classes of signaling pathways developed by prokaryotes. The external stimulus is perceived either by an internal receptor–transducer (left), or by a transmembrane histidine kinase that connects with a response regulator (center), or by an independent receptor associated to the histidine kinase (right). This scheme represents the basic taxonomy of bacterial signaling; the three different options imply very different information processing capabilities and metabolic costs.

The Signaling/Transcriptional Regulatory Networks, like the M. tuberculosis network, may be used to analyze the ability of Mycobacterium to perceive the host signals in different tissues and cell types, as well as adaptive responses that bacteria organizes against them. In that sense, to adequately study on processes of latency and reactivation using these networks would be very important. These networks will be useful to provide an overview of multiple functional aspects of this bacterium and to suggest new experiments.

Modeling Metabolites as "Metabolic Hardware"

In this we see the possibility of using metabolic networks as hardware in the study of the optimization of metabolic pathways as well as in the field of molecular and natural computing. We call to these bioinspired architectures as *metabolic hardware*. In particular, adopting as an example well known metabolic pathway of Krebs cycle, we introduce the methodology to translate the molecular structure or topology of their metabolic intermediates to a binary matrix, showing how sugars and other glycolytic molecules could be modeled as binary matrices as well as LED dot matrices. In prokaryotic cells and bacteria which lack mitochondria, the Krebs cycle is performed in the cytosol.

From a historical perspective one of the first procedures to translate the molecular topology to a matrix was introduced by Randic, taking an element a_{ij} the value 1 when the vertices are adjacent or 0 otherwise. Figure illustrates an example of this method for vitamin A or retinol. Our method assigns a 5-bit word to the functional groups of the molecule. For that purpose we define a table or Rosetta stone that includes the most frequent functional groups in metabolic intermediates, which were ordered by its redox potential (tendency of a functional group to acquire electrons).

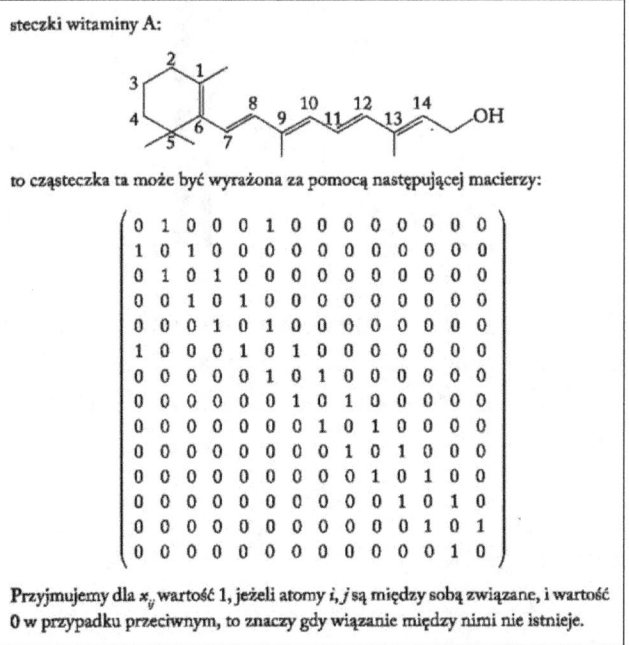

The molecule of vitamin A or retinol represented as a binary matrix.

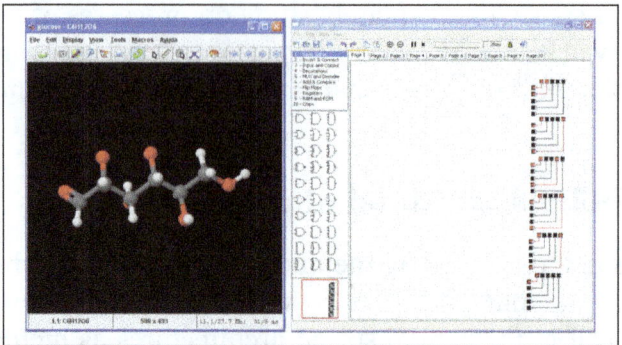

Glucose molecule (Left) and its hardware version as a matrix of
LEDs (Right) simulated with CEDAR Logic Simulator.

Rosetta stone for the hardware implementation of metabolic pathways.

Decimal	Binary	"Functional group"		Red-Ox scale
0	000 00			
1	000 01		Null values	
2	000 10			
3	000 11			
4	001 00			
5	001 01		Null values	
6	001 10			
7	001 11			
8	010 00	—CH₃		+Red
9	010 01	—CH₂—	Alkyle	- Ox
10	010 10	—CH—		
11	010 11	—C—		
12	011 00	=CH₂		
13	011 01	=CH—	Alkene	
14	011 10	=C—		
15	011 11	=C=		
16	100 00	—CH₂OH		
17	100 01	H—C—OH (S)	Alcohol	
18	100 10	HO—C—H (R)		
19	100 11	—C—OH		
20	101 00	—CH₂O—Ⓟ		
21	101 01	H—C—O—Ⓟ	Ester	
22	101 10	=C—O—Ⓟ		
23	101 11	—C—O—Ⓟ		
24	110 00	—C=O		
25	110 01	} Null values	Carbonyle	
26	110 10	}	(aldehide, ketone)	
27	110 11	—C—		
28	111 00	—C=O		
29	111 01	—C—O—Ⓢ	Carboxile, Ether, CO₂	
30	111 10	—C—O—Ⓟ		
31	111 11	CO₂		Red

Let S and P be two binary matrices which represent respectively the substrate S_m and product P_m of a biochemical reaction catalyzed by an enzyme E_m. Since in Krebs cycle all metabolites or metabolic intermediates are molecules with 4 or 5 carbon atoms, we defined and matrices respectively:

$$
C_4 = \begin{pmatrix}
a_{11} & a_{12} & a_{13} & a_{14} & a_{15} \\
a_{21} & a_{22} & a_{23} & a_{24} & a_{25} \\
a_{31} & a_{32} & a_{33} & a_{34} & a_{35} \\
a_{41} & a_{42} & a_{43} & a_{44} & a_{45}
\end{pmatrix}
$$

$$C_5 = \begin{pmatrix} a_{11} & a_{12} & a_{13} & a_{14} & a_{15} \\ a_{21} & a_{22} & a_{23} & a_{24} & a_{25} \\ a_{31} & a_{32} & a_{33} & a_{34} & a_{35} \\ a_{41} & a_{42} & a_{43} & a_{44} & a_{45} \\ a_{51} & a_{52} & a_{53} & a_{54} & a_{55} \end{pmatrix}$$

Note that given a value i, $(a_{i1}, a_{i2}, ..., a_{i5})$ is a row vector representing the functional group of the substrate s_{ij} or product p_{ij} molecules. Thus, each row in the matrices C_4 and C_5 represents a carbon atom in the molecule, having a total of 32 possible binary vectors from 00000 to 11111. Using as a criterion the redox potential vectors were classified from its most reduced (addition of hydrogen or the removal of oxygen) form or alkyl group to the most oxidized (addition of oxygen or the removal of hydrogen) or CO_2. However, since the metabolites of Krebs cycle are the result of assembling functional groups among a total of 22 combinations of carbon, then 10 binary vectors are without chemical meaning. In order to perform future simulation experiments, molecules of CO_2 and acetyl-CoA were represented as a row vector (7) and 2×5 matrix (8) shown below:

$$CO_2 = \begin{pmatrix} 1 & 1 & 1 & 1 & 1 \end{pmatrix}$$

$$\text{acetyl-CoA} = \begin{pmatrix} 1 & 1 & 1 & 0 & 1 \\ 0 & 1 & 0 & 0 & 0 \end{pmatrix}$$

Using this method the Krebs cycle was modeled as follows:

$$\begin{pmatrix} 1 & 1 & 1 & 0 & 0 \\ 0 & 1 & 0 & 0 & 1 \\ 1 & 0 & 0 & 1 & 1 \\ 0 & 1 & 0 & 0 & 1 \\ 1 & 1 & 1 & 0 & 0 \end{pmatrix} \rightarrow$$

$$\rightarrow \begin{pmatrix} 1 & 1 & 1 & 0 & 0 \\ 0 & 1 & 0 & 0 & 1 \\ 0 & 1 & 0 & 1 & 0 \\ 1 & 0 & 0 & 1 & 0 \\ 0 & 1 & 1 & 0 & 0 \end{pmatrix} \rightarrow \begin{pmatrix} 1 & 1 & 1 & 0 & 0 \\ 0 & 1 & 0 & 0 & 1 \\ 0 & 1 & 0 & 0 & 1 \\ 1 & 1 & 0 & 1 & 1 \\ 1 & 1 & 1 & 0 & 0 \end{pmatrix} \rightarrow \begin{pmatrix} 11100 \\ 01001 \\ 01001 \\ 11101 \end{pmatrix}$$

$$\leftarrow \begin{pmatrix} 1 & 1 & 1 & 0 & 0 \end{pmatrix}$$

where each matrix stands for one of the following metabolites: Citrate→ Iso-citrate→ α-Ketoglutarate→ Succinyl-CoA → Succinate → Fumarate → Malate → Oxalacetate.

$$\begin{pmatrix} 1 & 1 & 1 & 0 & 0 \\ 1 & 1 & 0 & 1 & 1 \\ 0 & 1 & 0 & 0 & 1 \\ 1 & 1 & 1 & 0 & 0 \end{pmatrix} \leftarrow \begin{pmatrix} 1 & 1 & 1 & 0 & 0 \\ 0 & 1 & 0 & 0 & 1 \\ 1 & 0 & 0 & 1 & 0 \\ 1 & 1 & 1 & 0 & 0 \end{pmatrix} \leftarrow \begin{pmatrix} 1 & 1 & 1 & 0 & 0 \\ 0 & 1 & 1 & 0 & 1 \\ 0 & 1 & 1 & 0 & 1 \\ 1 & 1 & 1 & 0 & 0 \end{pmatrix}$$

$$\leftarrow \begin{pmatrix} 1 & 1 & 1 & 0 & 0 \\ 0 & 1 & 0 & 0 & 1 \\ 0 & 1 & 0 & 0 & 1 \\ 1 & 1 & 1 & 0 & 0 \end{pmatrix}$$

Bacterial Evolution

At present, all methods in Evolutionary Computation (genetic algorithms, evolutive algorithms, genetic programming, etc.) are bioinspired by the fundamental principles of neo-Darwinism, and by a vertical gene transfer; that is to say, by a mechanism in which an organism receives genetic material from the ancestor from which it evolved. Indeed, most thinking in Evolutionary Computation focuses upon vertical gene transfer as well as upon crossover and mutation operations.

Bacteria are microscopic organisms whose single cells reproduce by means of a process of binary fission or of asexual reproduction, bearing a resemblance to John von Neumann's universal constructor. Thus, a bacterial population (or colony) evolves according to an evolutive algorithm similar to Dawkin's biomorphs, the cumulative selection of mutations powering their evolution. Bacteria, however, exhibit significant phenomena of genetic transfer and crossover between cells. This kind of mechanism belongs to a particular kind of genetic transfer known as horizontal gene transfer. Horizontal, lateral or cross-population gene transfer is any process in which an organism, i.e., a donor bacterium, transfers a genetic segment to another one, a recipient bacterium, which is not its offspring. In the realm of biology, whereas the scope of vertical gene transfer is the population, in horizontal gene transfer the scope is the biosphere. This particular mode of parasexuality between "relative bacteria" includes three genetic mechanisms: conjugation, transduction and transformation. Furthermore, microorganisms are very interesting individuals because they also exhibit "social interactions." We found how the inclusion of the "social life of microorganisms" into the genetic algorithm cycle, significantly improves the algorithm's performance.

Bacterial Conjugation

This topic describes a biologically inspired conjugation operator simulating a bacterial conjugation. Its usefulness is illustrated in a set of computer simulation experiments where including such operator into a genetic algorithm we were able to design an AM radio receiver. The attributes optimized by this algorithm include the main features of the electronic components of an AM radio circuit, as well as those of the radio enclosure designed to house the radio circuit.

A bacterial genetic algorithm is an evolutionary strategy based on bacterial conjugation and mutation. Starting with a random population of circular chromosomes reproduction, conjugation and mutation were simulated, obtaining new generations of equal size. The current bacterial algorithm uses homologous recombination after conjugation, a population size and a conjugation parameter, as well as a conjugation (or recombination) and mutation probabilities. Note that in the biological realm as well as in the simulations, the term population could be substituted by strain or colony and the linear chromosomes of a genetic algorithm are replaced by circular chromosomes. Our operator which includes the recombination between bacterial chromosomes assumes that donor bacterium is always Hfr. Two different conjugation operator versions have been defined. In both definitions since transfer of the donor bacterial chromosome is almost never complete, then the length of the strand transferred to the recipient cell has been simulated applying Monte Carlo's method and assuming DNA lengths exponentially distributed:

$$l = -\frac{1}{\alpha} \ln(U)$$

being U a random number and α the conjugation parameter. The conjugation parameter summarizes all the relevant factors affecting the length l value. One of the most relevant factors affecting value is the temperature promoting the agitation of the bacteria, disrupting conjugation before the entire chromosome can be transferred.

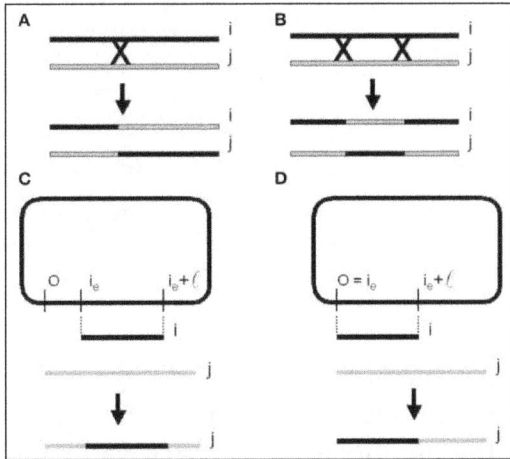

Homologous recombination or cross over mechanisms. (A) One-point re-combination. (B) Two-points recombination. Bacterial conjugation and recombination. (C) With a random point (CORP). (D) With a fixed point (COFP) on the donor chromosome.

In bacteria, crossing over involves the aligning of the donor chromosome segment with its homologous segment on the recipient bacterial chromosome. Next, a break occurs at a point origin and an end point of the recipient chromosome, removing and replacing the segment with corresponding homologous genes from the segment of the donor chromosome. The described steps are repeated several times, thus a number of times equal to the bacteria population size. The efficiency of the bacterial conjugation operator has been illustrated designing an AM radio receiver with a genetic algorithm based on this operator.

We show the bacterial chromosome coding for the main features of the radio receiver and in figure a representative performance graph (average fitness per generation) of the experiments where simulated bacterial colonies evolved searching for the optimized circuit and enclosure.

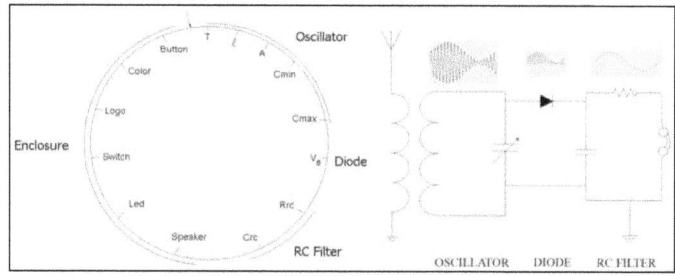

Bacterial chromosome (left) with 14 genes codifying for the main characteristics of an AM radio receiver (right).

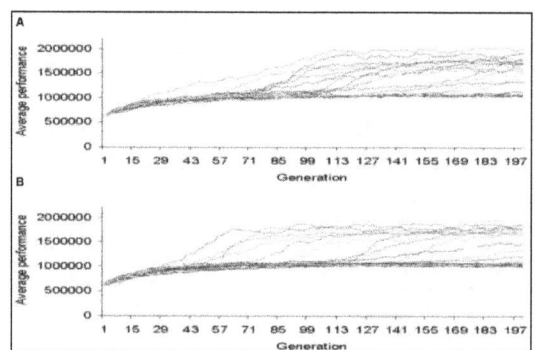

Performance graph where bacterial colonies evolved searching for an optimized AM radio receiver. (A) CORP. (B) COFP.

Bacterial Transduction

In Nature, microorganisms such as bacteria and viruses share a long and common evolutionary relationship. This relationship is mainly promoted by bacteriophages (or phages), a kind of virus that multiplies inside bacteria by making use of the bacterial biosynthetic machinery. Some bacteriophages are capable of moving bacterial DNA (the "bacterial chromosome") from one bacterium to another. This process is known as transduction. When bacteriophages infect a bacterial cell, their normal mode of reproduction makes use of the bacterium's replication machinery, making numerous copies of its own viral genetic material (i.e., DNA or RNA). The nucleic acid copies (or chromosome segments) are then promptly packaged into newly synthesized copies of bacteriophage virions. Generalized transduction occurs when "any part" of the bacterial chromosome (rather than viral DNA) hitchhikes into the virus (i.e., T4 phages in *E. coli* bacterium). However, when only "specific genes" or certain special "segments" of the bacterial chromosome can be transduced, such a mistake is known as specialized transduction (i.e., λ phages in *E. coli* bacterium).

In transduction, transference of chromosome segments between bacterial populations or colonies is very different from migration (the occasional exchange of individuals). Migration and transduction could bear a resemblance, but only when transduction involves the complete chromosome transference between bacterial populations. Furthermore, this kind of transference is a highly unlikely event in bacteria, transduction of chromosome segments taking place in these microorganisms.

In this topic, we model and simulate the two kinds of transduction operations examining the possible role and usefulness of this genetic mechanism in genetic algorithms. We introduced a bacterial conjugation operator showing its utility by designing an AM radio receiver. Conjugation is one of the key genetic mechanisms of horizontal gene transfer between bacteria. A genetic algorithm including transduction as PETRI (Promoting Evolution Through Reiterated Infection). We investigated the transfer of genes and chromosomes among sub-populations with a simulated "bacteriophage." In the model we consider a structured population divided among several sub-populations or "bacterial colonies," bearing a resemblance with coarse-grain distributed genetic algorithms. Each sub-population is represented as a Petri dish (a glass or plastic cylindrical dish used to culture microorganisms). It should be noted, however, that even when we divide a population into sub-populations, the proposed algorithm is sequential. Thus, the algorithm is not a distributed one, since we used a mono-processor computer and the algorithm was not parallelized. Moreover, the migration mechanism is synchronous, as gene and chromosome transferences were both between sub-populations and during the same generation. Therefore, our approach could be related with those models of Cellular Genetic Algorithms (cGA) adopted also for mono-processor machines, with no relation to parallelism at all. In our model, we assumed that bacteria are capable of displaying crossover through conjugation, instead of performing one-point or two-point recombination. Moreover, we assume that no vertical gene transfer mechanism is present in bacterial populations.

With the aim of studying the performance of the transduction operator, we used different optimization problems. Experiments conducted in the presence of transduction were compared with control experiments, performed in the absence of transduction. Similarly, we compared the transduction performance under the three types of crossover: conjugation, one-point or two-point recombination. We are interested in the study of genetic algorithms based on horizontal gene transfer mechanisms, mainly conjugation and transduction operations. It is important to note that even when conjugation and transduction are both horizontal gene transfer mechanisms, there are some relevant differences between both. In the first place, whereas conjugation involves two bacteria from the same population, the bacteria involved in transduction can belong to different populations. As a consequence, conjugation is a genetic mechanism of horizontal gene transfer within a population, whereas transduction is a genetic mechanism of horizontal gene transfer between populations. Secondly, in conjugation, the length of the transferred genetic segment is variable, whereas in transduction, the transferred segment length is always constant.

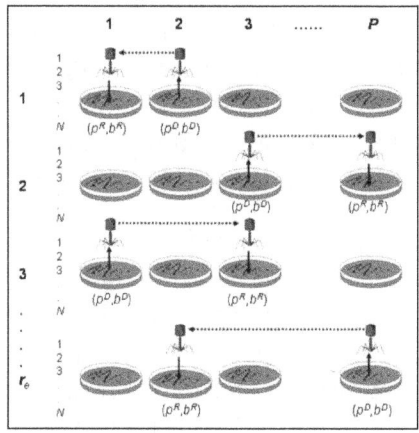

Transduction experiment: The figure shows transduction from donor Petri dish (p^D) and bacterium (b^D) to recipient Petri dish (p^R) and bacterium (b^R). In the figure, P is the total number of Petri dishes (or sub-populations), N is the number of bacteria (or population size) per Petri dish and r_e the number of experimental replicates.

Let b be a chromosome (i.e., bacterium; 1, ..., j, ..., N) and p a sub-population (i.e., Petri dish; 1,..., i,..., P); then a transduction operation is defined as follows: transduction is the transfer of genetic material from a Petri dish and bacterium donors (p^D, b^D) to a Petri dish and bacterium recipients (p^R, b^R). When the transference involves a chromosome segment, the result is a recombinant chromosome in the recipient Petri dish p^R. However, the transference of a complete chromosome results in the substitution of one chromosome of the recipient Petri dish p^R with the transferred one. It is important to note that "bacterium" and "Petri dish" terms are used as "chromosome" and "sub-population" synonyms, respectively. Transduction requires the selection of the Petri dish and bacterium donors (p^D, b^D), as well as the Petri dish and bacterium recipients (p^R, b^R). In the reference we describe how transduction was conducted.

The current PETRI algorithm uses a population size of N, performing r_e replicates, with P being the total number of Petri dishes or sub-populations. Thus, we performed a number of r_e. P trials of each simulation experiment. The algorithm cycles through epochs, searching for an optimum solution until a maximum of G generations is reached. Once (p^D, b^D) and (p^R, b^R) are selected, only one "bacteriophage" is assumed to participate during each transduction event. The PETRI algorithm is summarized in the following pseudocode description:

/* PETRI: Genetic Algorithm with Transduction */

1. t:=0;

2. Initialization: Generate P Petri dishes (or sub-populations)

with N random bacteria (or chromosomes).

3. WHILE not stop condition DO

/* Genetic Algorithm */

(3.1) FOR each P Petri dish DO

Evaluation of chromosomes

Selection

Conjugation or Crossover (one-point, two-point)

Mutation

(3.2) END FOR

/* End of Genetic Algorithm */

4. Transduction: (pD, bD) (pR, bR)

5. t:=t+1;

6. END WHILE;

/* End of PETRI */

We studied the performance of the simulated transduction by considering three optimization problems that are described in sufficient detail in Perales-Graván et al. The first problem uses a benchmark function, the second one is the 0/1 knapsack problem, and finally we illustrated the usefulness of transduction in the problem of designing an AM radio receiver.

IMPORTANCE OF BACTERIAL CULTURE TO FOOD MICROBIOLOGY

Culture-based and genomics methods provide different insights into the nature and behavior of bacteria. Maximizing the usefulness of both approaches requires recognizing their limitations and employing them appropriately. Genomic analysis excels at identifying bacteria and establishing the relatedness of isolates. Culture-based methods remain necessary for detection and enumeration, to determine viability, and to validate phenotype predictions made on the bias of genomic analysis. The purpose of this short paper is to discuss the application of culture-based analysis and genomics to the questions food microbiologists routinely need to ask regarding bacteria to ensure the safety of food and its economic production and distribution. To address these issues appropriate tools are required for the detection and enumeration of specific bacterial populations and the characterization of isolates for, identification, phylogenetics, and phenotype prediction.

Genomics, the study of the encoding, structure and function of genetic information, can be considered to have emerged as a recognized discipline with the initial publication of the eponymous journal in 1987. Initial efforts to sequence the complete genome of organisms required heroic investments of ingenuity and resources. The first prokaryote sequence, that of *Haemophilus influenzae*, was published in 1995 and the first eukaryote sequence, *Saccharomyces cerevisiae*, in 1996. The first human genome was completed in 2004, following a multinational effort with an estimated cost of 3 billion dollars. The return on these investments has been techniques and tools which have dramatically lowered the time and cost of sequencing and the time and expertise required for analysis. For bacteriologists, genomic analysis is becoming a routine tool, with whole genome sequencing (WGS) available affordably, in a matter of days and with analysis supported by online platforms.

The foundation of bacteriology as an experimental science was the development in the 19th century of cultural techniques using solid or liquid media. Culture allowed viable bacteria to be detected and isolated, facilitated the observation of metabolic activity, and provided biomass for further analysis. As genomic analysis of bacteria becomes available as a routine tool, there is a risk of a perception developing that genomic analysis of bacteria is superior or may supersede bacterial cultural methods due to the of speed analysis and quantity of data produced. The purpose of this paper is to discuss the questions that food microbiologists routinely consider regarding bacteria, and to examine the application of cultural and genomic approaches to answering them. In food microbiology bacteria as infectious pathogens and toxin producers are a safety concern. Bacteria are also an economic concern, as their metabolic activity may enhance the economic value of foods or negatively impact it by altering sensory qualities or nutritional content.

Consideration of the strengths and limitations of cultural and genomic based methods of analysis of bacteria indicates that they provide different insights into the nature and behavior of bacteria, with the former revealing phenotypic characteristics and behavior and the later genotypic information. Neither type of information is superior to the other, but rather they should be viewed as specialized and complementary tools which are suited to answering different experimental questions.

The Detection and Enumeration of Bacteria in Foods

The most commonly used forms of bacteriological analysis in food microbiology are detection and enumeration. The presence of specific bacteria and their concentration must be determined, to assess and control safety hazards, the potential for spoilage or to ensure correct product characteristics. The bacteria of interest to food microbiology can be divided into infectious agents, causes of foodborne intoxication, spoilage, and processing aids. Metabolic activity of a bacterium may be considered as causing spoilage or as a processing aid depending upon the desirability of the changes that result.

Table: Examples of bacteria of concern to food microbiology.

Foodborne infectious agents	Foodborne intoxicants	Spoilage	Processing
Brucella	Bacillus cereus	Acinetobacter	Lactic Acid
Campylobacter	Clostridum botulinum	Alcaligenes	Bacteria
Clostridium botulinum	Clostridium perfringens	Bacillus	(Lactobacillus,
Clostridium perfringens	Staphylococcus aureus	Brochothrix	Lactococcus,
Cronobacter		thermosphacta	Pediococcus,
pathogenic Escherichia coli		Clostridium	Leuconostoc,
Shigella		Cornebacterium	Streptococcus)
Salmonella enterica		Enterobacteriaceae	
Yersinia enteroclitica		Erwinia carotovora	
Listeria monocytogenes		Lactic Acid Bacteria	
Mycobacterium		Moraxellaceae	
Vibrio		Pseudomonas	
		Shewanella	
		putrefaciens	
		Vibrio	

Detection of specific types of bacteria can be achieved by cultural isolation, or by indicators such as biomolecules specific to the organism (e.g., nucleic acid sequences, antigens, toxins) or products of metabolism (e.g., gas, acid, substrates with chromogenic products). For enumeration cell-concentration can be estimated by partitioning the sample upon a solid surface (e.g., agar media, membrane), between liquid aliquots (e.g., most probable number) or through direct or indirect measures of biomass (e.g., optical density, Limulus amebocyte lysate assay).

Culture independent diagnostic platforms for infectious agents have been successfully commercialized in the health care sector and the potential for speed and automation has stimulated interest in the application of similar systems for food analysis. Platforms developed for health care cannot be easily adopted for use in food microbiology as analysis is significantly more challenging. Food microbiology samples are considerably more variable in type, heterogeneous in composition and the concentrations of target bacteria can be much lower. Also with the exception of stools, body fluids and tissue samples can normally be expected to contain a negligible microbiota.

The presence of non-target microbiota is a particular problem when testing for pathogens as closely related non-pathogens may result in false positives, which can have serious implications for producers. For example, the US Department of Agriculture requires testing of raw ground beef components for shiga toxin-producing *E. coli* (STEC) which possess three traits: the virulence genes *stx* and *eae*, and six O-types considered of high risk. *E. coli* strains which possess one or two of these traits are considered of low risk and do not require the same regulatory response. The three traits, however, are not genetically linked and may be present within the sample in multiple different organisms.

Genomic technologies are considered appealing for culture-independent detection as reliability can be provided by the parallel detection of multiple genes or their transcription products. Though not currently possible, in principle, WGS using sufficiently long reads could detect and confirm the presence of the complete genome of multiple target organisms in a complex mix of DNA. In spite of accelerating advances, however, genomic technologies are not suitable for addressing two fundamental challenges in detection and enumeration; sensitivity and the determination of viability.

The Sensitivity of Cultural Methods for Bacterial Detection

The sensitivity or limit of detection (LOD) of methods of analysis for bacterial cells is the minimum concentration of cells that can be detected. Analysis for the presence of bacteria that cause foodborne intoxication, spoilage or serve as production aids does not generally require limits of detection below 100 CFU/g or ml. Spoilage and processing bacteria do not impact the quality of a product until they exceed a significant concentration, for example spoilage of red meats by *Pseudomonads* becomes apparent above 6 log CFU/cm^2 . Bacteria which causes intoxication need to reach relatively high concentrations in foods before significant toxin production occurs. For *C. botulinum* the threshold is 3 log CFU/g and for *B. cereus* and *S. aureus* cell concentration must exceed 5 log CFU/g. However, some infectious agents have infectious doses estimated in the range of 10–100 cells, and the concentration of pathogen cells in outbreak associated products may be below 1 cell per 25 g. Thus, regulatory compliance testing of foods for infectious bacterial pathogens requires LODs approaching 1 cell per analytical unit, with analytical units of 10 g to 325 g depending on the specific pathogen and food.

Without enrichment, no existing technologies can approach this sensitivity. Whether or not analysis is based on detection of cells or biomolecules, the target of analysis needs to be separated from the surrounding complex organic matrix, without loss of the target by adherence to the analytical apparatus. The relatively large analytical unit sizes (10 g to 325 g) make it impractical to assay anything other than a smaller aliquot of the analytic unit (0.1 to 1 ml) and many foods are composed of solids, gels and suspensions, with consequent heterogeneous distribution of bacterial cells. A method of analysis which is dependent upon the probability of the target being present in an aliquot of the analytical sample is inherently unreliable.

Cultural enrichment resolves the challenge of high sensitivity bacterial detection by amplifying the analytic target to raise the concentration and distribute it homogenously through an aqueous suspension. This ensures that detection is no longer a probabilistic process and sample handling is greatly eased. The only limitation is that the minimum time required for sample analysis is determined by the enrichment period. This will be determined by the time required for cells to begin replication (repair injury and exit lag phase) and the time required to reach the LOD of the method of analysis (growth rate). The enrichment period could be reduced by a concentration process once the analytic target is homogenously distributed in the suspension. However, the time and resources required for concentration may make this less efficient than extending the enrichment period.

Determination of Bacterial Viability

For the bacteriological analysis of foods, it is highly desirable that the method of analysis does not confound viable cells (cells with the potential to replicate), with non-viable cells or cell debris. Foods often undergo processing steps which impact bacterial survival: the resulting population of cells may include viable cells, cells that can replicate following repair (reversibly injured) and cells which can not replicate but retain metabolic activity (irreversibly injured). Cells of infectious agents that cannot replicate pose no threat to health. Non-replicating toxin producers only pose a threat if their concentration is already high enough to present a risk. Similarly, the presence of non-replicating cells of spoilage and processing bacteria are of no relevance to food quality.

Analytical methods which detect the presence of biomolecules, such as DNA, RNA, or proteins, cannot determine whether those biomolecules represent viable cells or not. Cell replication is a complex process, in which multiple regulatory mechanisms must coordinate the synthesis and localization of a vast array of structural and functional molecules. The failure of the cell to complete any essential function can stall cell growth and division. Confirmation of the presence of any single essential cell component does not exclude other deficiencies that would inhibit cell replication. Thus, viability can only be determined by two methods, determination of the presence and functionality of all the molecules required, or simply waiting for the cell to exit lag phase and allowing replication to occur.

Assessment of viability may be further complicated by the potential for vegetative bacterial cells, including some foodborne pathogens, to enter into an alternate physiological state, viable-but-nonculturable (VBNC). The VBNC state can be triggered by a variety of physiochemical stresses, with cells ceasing replication but continuing metabolic activity. Experimentally distinguishing VNBC states from injury is experimentally complicated, but since by definition VNBC cells can resume replication following exposure to an appropriate resuscitation stimulus identifying the correct stimulus for resuscitation rather than abandoning culture appears a more productive response.

Charactering Bacteria

Bacterial isolates are characterized to confirm identity, to establish relationships between isolates and to understand the behavior (phenotype) of the bacterium. Though a variety of cultural and molecular methods for characterization are available genomic analysis is superseding them. Genomics may have clear superiority over other approaches in identifying and subtyping isolates, but its application to predict phenotype is much less reliable.

Identification and Subtyping

Genomic analysis to identify isolates to the genus or species level by 16S rRNA sequence and the detection of specific gene markers by polymerase chain reaction based methods is more reliable and has greater discrimination than phenotypic methods, particularly for identification below the species level. WGS analysis is superseding other approaches, as though complex computational analysis is required, a single wet lab process can provide information on the presence of multiple gene markers and phylogenetic relationship. Additionally, sequence data can be retrospectively analyzed for additional markers or potential relationships.

Establishing phylogenetic relationships for bacterial isolates below the species level can link isolates from clinical, food and environmental samples, for outbreak identification and source tracking. Single polynucleotide polymorphism (SNP) analysis of WGS data can provide unprecedented discrimination between isolates. Methods for the prediction from WGS data of established genetic subtyping such as pulsed field gel electrophoresis, multilocus sequence typing and multiple-locus variable number tandem repeat analysis are being developed, but the accuracy can be compromised by short read lengths. Whether subtyping of isolates is by SNP or alternatives, the only limitation on the relating isolates is the comprehensiveness of WGS and accompanying metadata available. For example, WGS data has been used to deduce the geographical origin of isolates.

It should be noted that the application of WGS data for isolate identification, subtyping and source tracking in the context of public health, regulatory, and commercial decision making is very recent and standards for analysis and interpretation have yet to be established. Reported results are dependent upon the sequencing platform used (length of reads, error rate, genome coverage), and the data analysis pipeline. Data interpretation may be significantly affected by choices such as nucleotide identification algorithms, assembly method and whether the shared or core genome of isolates is compared. The need to address these issues by establishing analytical standards is recognized, but there will be a transition period until a consensus on standards emerges.

Predicting Bacterial Phenotype

The complementary nature of genomic and phenotypic data is apparent when attempting to understand and predict bacterial behavior. There is wide range of bacterial behavior of interest to the food microbiologist. These include the potential for survival and replication during food production and distribution, spoilage potential, and the hazard potential of pathogens. Genomic analysis allows researchers to rapidly detect known genes, putative genes, and other defined features of the genome. However, relating genotype to phenotype with accuracy is highly challenging. Many phenotypic characteristics are the product of multiple genes and their regulatory systems. Current knowledge of any but a handful of biological regulatory systems is far from perfect. The same phenotype may result from multiple mechanisms with differing genotypes. The phenotype may also be dependent upon interactions with other organisms. Genetically homogenous cells may be phenotypically heterogeneous, as observed in persister cells and biofilms. Epigenetic inheritance, DNA methylation and small RNAs have been identified as playing a role in determining bacterial phenotype, but the mechanisms are not well understood. The significance of other potential epigenetic mechanisms such as prions, self-sustaining metabolic loops, and structural templating of membranes in bacteria is unknown.

The challenges and the opportunities presented by predicting phenotype from genotype are illustrated by the prediction of antimicrobial resistance (AMR) from WGS data. Many WGS analysis platforms provide output of AMR associated genes, but the utility of this information is questionable. A 2016 review of the potential for AMR prediction by WGS conducted for the European Committee on Antimicrobial Susceptibility Testing concluded that, for the purposes of informing clinical decisions, "The published evidence for using WGS as a tool to infer antimicrobial susceptibility accurately is currently either poor or non-existent". Though prediction accuracy may be limited, the ability to rapidly screen large populations of strains for a potential phenotype is still useful. used WGS data to determine whether a specific STEC strain possessed AMR genes that were relatively uncommon among *E. coli* and determined the presence of trimethoprim resistance genes. Trimethoprim resistance was confirmed experimentally and the addition of trimethoprim to enrichment broth was demonstrated to aid isolation. In an outbreak investigation, this approach could be used in the analysis of foods for strains previously isolated from patients.

The purpose of discussing the limitations of genomics is not to imply that the application of genomics to answer these questions is inappropriate. When complemented with phenotypic data and studies of physiological mechanisms, genomic data is a powerful tool to improve our understanding, but the challenges in relating genomic data to phenotype must be recognized. Decision making related to food safety or food processing should not be made solely on the basis of genomic data, but needs to be supported by phenotypic data, which in turn require culture. When grounded in phenotypic data, genomic data has the potential to enhance culture methods or to develop culture method for previously unculturable organisms. As databases of WGS data expand, genomic analysis can be used to rapidly screen large populations for strains that potentially possess desired phenotypes, or to select experimental strains that are representative of a larger population.

Genomic technologies are tools with enormous potential for increasing our understanding of bacteria and solving practical problems in food microbiology, but like any tools the benefits and costs are dependent upon how we choose to employ them. Medical microbiology is faced with a set of unnecessary challenges due to the trend of abandoning cultural isolation for culture-independent diagnostic testing. At the clinical level this presents difficulties distinguishing viable from non-viable organisms and in data interpretation when multiple organisms are present. At the public health level this results in the inability to collect epidemiological data such as, subtype and AMR, and prohibits further characterization and research. Food microbiologists should learn from this example and consider how to maximize the benefits without losing the advantages of alternate technologies. Genomic analysis is becoming the standard method for the identification and phylogenetics of bacteria, but culture remains necessary to achieve the required sensitivity of detection and enumeration and to determine viability. Just as crucially, culture is needed to provide the isolates with which to conduct experiments to test hypothesizes generated from genomic data. When phenotype is predicted from genomic data we are creating a model of a biological system and the great value of such models as noted by Jeremy is to reveal the limitations of our understanding.

References

- Medical-uses-for-bacteria, opinion: pharmaceutical-journal.com, Retrieved 15 May, 2019

- Hsu, ronald. "fecal microbiota transplantation (fmt), bacteriotherapy". American college of gastronomy. Retrieved 3 june2018

- Bacteriology: bacteriology.net, Retrieved 23 June, 2019

- Million, m; lagier, jc; yahav, d; paul, m (2013). "gut bacterial microbiota and obesity". Clinical microbiology and infection. 19 (4): 305–313. Doi:10.1111/1469-0691.12172. Pmid 23452229

- Cryan, jf; dinan, tg (2012). "mind altering microorganisms: the impact of the gut microbiota on brain and behavior". Nature reviews neuroscience. 13 (10): 701–712. Doi:10.1038/nrn3346. Pmid 22968153

- Kang, dw; adams, jb; gregory, ac (2017). "microbiota transfer therapy alters gut ecosystem and improves gastrointestinal and autism symptoms: an open label study". Microbiome. 5 (1): 10. Doi:10.1186/s40168-016-0225-7. Pmc 5264285. Pmid 28122648

Permissions

We would like to thank the editorial team for lending their expertise to make the book truly unique. They have played a crucial role in the development of this book. Without their invaluable contributions this book wouldn't have been possible. They have made vital efforts to compile up to date information on the varied aspects of this subject to make this book a valuable addition to the collection of many professionals and students.

This book was conceptualized with the vision of imparting up-to-date and integrated information in this field. To ensure the same, a matchless editorial board was set up. Every individual on the board went through rigorous rounds of assessment to prove their worth. After which they invested a large part of their time researching and compiling the most relevant data for our readers.

The editorial board has been involved in producing this book since its inception. They have spent rigorous hours researching and exploring the diverse topics which have resulted in the successful publishing of this book. They have passed on their knowledge of decades through this book. To expedite this challenging task, the publisher supported the team at every step. A small team of assistant editors was also appointed to further simplify the editing procedure and attain best results for the readers.

Apart from the editorial board, the designing team has also invested a significant amount of their time in understanding the subject and creating the most relevant covers. They scrutinized every image to scout for the most suitable representation of the subject and create an appropriate cover for the book.

The publishing team has been an ardent support to the editorial, designing and production team. Their endless efforts to recruit the best for this project, has resulted in the accomplishment of this book. They are a veteran in the field of academics and their pool of knowledge is as vast as their experience in printing. Their expertise and guidance has proved useful at every step. Their uncompromising quality standards have made this book an exceptional effort. Their encouragement from time to time has been an inspiration for everyone.

The publisher and the editorial board hope that this book will prove to be a valuable piece of knowledge for students, practitioners and scholars across the globe.

Index

www.ingramcontent.com/pod-product-compliance
Lightning Source LLC
Chambersburg PA
CBHW080403190526
45161CB00003B/114